AF738370

Corporate Sustainability in International Comparison

ECO-EFFICIENCY IN INDUSTRY AND SCIENCE

VOLUME 31

Series Editor: Arnold Tukker, TNO-STB, Delft, The Netherlands

Editorial Advisory Board:

Martin Charter, *Centre for Sustainable Design, The Surrey Institute of Art & Design, Farnham, United Kingdom*
John Ehrenfeld, *International Society for Industrial Ecology, New Haven, U.S.A.*
Gjalt Huppes, *Centre of Environmental Science, Leiden University, Leiden, The Netherlands*
Reid Lifset, *Yale University School of Forestry and Environmental Studies, New Haven, U.S.A.*
Theo de Bruijn, *Center for Clean Technology and Environmental Policy (CSTM), University of Twente, Enschede, The Netherlands*

For further volumes:
http://www.springer.com/series/5887

Stefan Schaltegger • Sarah Elena Windolph
Dorli Harms • Jacob Hörisch
Editors

Corporate Sustainability in International Comparison

State of Practice, Opportunities and Challenges

EMAN

SM Centre for Sustainability Management

Springer

Editors
Stefan Schaltegger
Sarah Elena Windolph
Dorli Harms
Jacob Hörisch
Centre for Sustainability Management (CSM)
Leuphana University
Lüneburg, Germany

ISSN 1389-6970
ISBN 978-3-319-06226-6 ISBN 978-3-319-06227-3 (eBook)
DOI 10.1007/978-3-319-06227-3
Springer Cham Heidelberg New York Dordrecht London

Library of Congress Control Number: 2014942330

© Springer International Publishing Switzerland 2014
This work is subject to copyright. All rights are reserved by the Publisher, whether the whole or part of the material is concerned, specifically the rights of translation, reprinting, reuse of illustrations, recitation, broadcasting, reproduction on microfilms or in any other physical way, and transmission or information storage and retrieval, electronic adaptation, computer software, or by similar or dissimilar methodology now known or hereafter developed. Exempted from this legal reservation are brief excerpts in connection with reviews or scholarly analysis or material supplied specifically for the purpose of being entered and executed on a computer system, for exclusive use by the purchaser of the work. Duplication of this publication or parts thereof is permitted only under the provisions of the Copyright Law of the Publisher's location, in its current version, and permission for use must always be obtained from Springer. Permissions for use may be obtained through RightsLink at the Copyright Clearance Center. Violations are liable to prosecution under the respective Copyright Law.
The use of general descriptive names, registered names, trademarks, service marks, etc. in this publication does not imply, even in the absence of a specific statement, that such names are exempt from the relevant protective laws and regulations and therefore free for general use.
While the advice and information in this book are believed to be true and accurate at the date of publication, neither the authors nor the editors nor the publisher can accept any legal responsibility for any errors or omissions that may be made. The publisher makes no warranty, express or implied, with respect to the material contained herein.

Cover logo captions: © Environmental and Sustainability Management Accounting Network
© Centre for Sustainability Management

Printed on acid-free paper

Springer is part of Springer Science+Business Media (www.springer.com)

Foreword

Sustainability is a global challenge that requires a global response. Environmental pollution, climate change, deforestation, declining biodiversity, and ecosystem degradation are problems that often cross international boundaries. The international community needs to come together to solve them at a global level. For this to be achieved we first need to understand the state of these problems and how companies are dealing with sustainability issues.

I have known and admired the work of Stefan Schaltegger for many years. I visited his Centre for Sustainability Management in Lüneberg for the "Sustainability Summit" in March 2012 and got to know both his local team and his wide network of international contacts. I cannot think of a better group to pull together an international collaborative project of this scope. This volume covers an impressive array of international corporate sustainability issues. It examines companies' intentions to pursue sustainability, their integration of sustainability into their core business, and implementation of sustainability management measures. These issues are uniformly tracked in many nations in America, Asia, Australia and Europe. The book is based on surveys of sustainability managers in large companies. This empirical base grounds its findings in real world practices, and offers an International Corporate Sustainability Barometer.

By focusing on individual country challenges this study documents the variations, range and differences in corporate sustainability problems and practices in different locations and industries. No single solution will work in different countries. Collectively the practices and solutions cover a dizzyingly large array of issues including water, soil, biodiversity, environmental pollution, geological structures, stakeholders (business partners, investors, consumers, suppliers), role of trade association and communities, top management commitment, legitimation pressures, business strategies, business models, government influence, legislation, internal differences between units, social issues, corporate social responsibility, communications, employee participation, employee tools and freedom, skepticism about benefits, ... as the reader will discover this list is actually longer. The most heartening aspect of these diverse variables is that they hold out promise of being applicable in any company. In this sense the International Corporate Sustainability Barometer will be a very useful tool for diagnosing areas for companies to work on to improve their sustainability performance.

Another powerful finding of this study is that sustainability management not only improves the triple bottom line, it also secures legitimacy claims for companies in most countries – a compelling reason for all international companies to vigorously engage with sustainability. And this book is a good place to start understanding the international state of the art.

This is a unique volume and it is very timely. It fills a much-needed gap in our understanding of internationally comparative corporate sustainability practices.

Montreal, QC, Canada Paul Shrivastava

Foreword

On behalf of ABIS – The Academy of Business in Society – I am honoured to have been asked by Professor Stefan Schaltegger and colleagues to contribute a foreword to *Corporate Sustainability in International Comparison*. It is the latest addition to the impressive body of work generated by the Centre for Sustainability Management in its first decade of existence. Like its predecessors, it achieves the twin goals of enriching knowledge and informing practice – something which (regrettably) cannot be said about a great deal of outputs from management research today.

This publication is in my view both timely and highly material to a wider debate about the future role of business in society. Nowadays, and especially since the financial crisis, *Corporate Sustainability* is gaining more currency and traction within the global business community.

When one asks managers what they mean with this term, they refer to a 'lasting business', built on 'long-term value creation', inclusive of 'stakeholder interests', supported by 'trust and legitimacy', with 'respected and reliable products and services', with 'high standards in production, supply and distribution chains', 'responsive to societal expectations and context requirements', and a 'stable financial model' which is 'resilient to stock market fluctuations and short-term financial concerns'.

In 2012 ABIS celebrated its own 10 year anniversary. In doing so, we surveyed global CEOs and Deans of business schools on how they perceived the transition from corporate responsibility to corporate sustainability (or sustainable business) in their own organisations. We also asked them what they felt were the critical skills and knowledge-based capabilities that would be needed in the turbulent post-crisis context.

One of the business leaders we interviewed responded as follows: "It is not enough to manage the business in the traditional way; it is equally important to manage the context requirements in which we operate. That might be a regional or country-specific context (like China), but also the context of the value chain from upstream to downstream. We need to define how stakeholders should be involved at each point in this value chain to create genuine shared value."

This perspective highlights the importance to business of the kind of study which underpins *Corporate Sustainability in International Comparison*. Global companies are increasingly challenged to tailor their business strategy and sustainability-related policies to regional or local contexts, avoiding a "one size fits all" approach in favour of flexible, smart, inclusive practice on the ground. By extension, there is a growing interest among MNCs to benchmark their management approaches and performance against rivals and peers in different parts of the world.

Against this backdrop, *Corporate Sustainability in International Comparison* deserves to attract a wide and diverse readership, both within industry but also among other enterprise stakeholders. It will hopefully prompt reflection and debate within many companies on how to enhance their management of global sustainability challenges, with the resultant benefits for their shareholders and diverse societies within which they operate. Once again, I congratulate Professor Schaltegger and his colleagues for their notable achievement.

Brussels, Belgium Simon Pickard

Preface

The beginning of the twenty-first century is characterised by growing knowledge of sustainability topics. Various initiatives like the UN Decade for Sustainability in Education have increased the awareness of pressing sustainability challenges such as climate change, biodiversity loss, poverty, or equal opportunities.

While a broad agreement exists among researchers of multiple disciplines that sustainable development of society also requires a sustainable development of companies, little is known about sustainability management practices internationally. This edited volume compiles international survey data collected and analysed together with renowned academic partner institutions from all over the world. The International Corporate Sustainability Barometer (ICSB) survey addresses important questions of the *intention* of large companies to deal with corporate sustainability, the *integration* of sustainability issues and management into the organisation and operational management as well as the actual *implementation* in corporate practice. We investigate how corporate sustainability practices differ in various economically developed countries and which future paths can be taken to increase the contribution of large companies to sustainable development worldwide.

To tackle these questions, Part I of this volume describes the approach and summarizes the overall results. Chapter 1 introduces the main ideas and structure of the book whereas Chap. 2 highlights important research gaps and the methodology. An overview of the aggregate results of the ICSB survey is provided in Chap. 3. Part II presents the specific findings for each country in alphabetical order. Chapters 4, 5, 6, 7, 8, 9, 10, 11, 12, 13 and 14 offer detailed analyses on the state of the art of corporate sustainability management in Australia, Belgium, France, Germany, Hungary, Japan, Korea, Spain, Switzerland, the UK and the USA. Part III concludes with a comparative analysis and highlights overall patterns of the international results. Most strikingly, the book reveals surprisingly broad similarities among the sustainability management practices of large companies in different economically developed countries all over the world.

We are grateful for the opportunity to realise this international project with a great team of researchers. United, we hope that the twenty-first century will not only be kept in mind as a century of growing sustainability awareness and increasing sustainability problems but also as a century of transition towards effective and beneficial corporate sustainability management practices. It is our deep hope that this edited volume supports the sustainability transition by providing researchers and practitioners alike with useful insights and inspirations.

Lüneburg, Germany

Stefan Schaltegger
Sarah Elena Windolph
Dorli Harms
Jacob Hörisch

Contents

Contributors

Igor Álvarez Basque Country University, San Sebastián, Spain

Amel Ben Rhouma CEDAG Management EA-1516 IUT, Department of Management and Administration, University Paris-Descartes, Paris, France

Roger Burritt Faculty of Business and Economics, Department of Accounting and Corporate Governance, Macquarie University, Sydney, Australia

Amanda Carter Centre for Accounting, Governance and Sustainability, School of Commerce, University of South Australia, Adelaide, Australia

Nathalie Crutzen HEC-Management School of the University of Liege, Liege, Belgium

Maria Csutora Sustainability Indicators Research Centre, Corvinus University of Budapest, Budapest, Hungary

Claus-Heinrich Daub School of Business, Institute of Management, University of Applied Sciences and Arts Northwestern Switzerland, Windisch, Switzerland

Claude Francoeur Department of Accounting Studies, HEC Montreal, Montreal, Canada

Biswaraj Ghosh Nottingham Business School, Nottingham Trent University, Nottingham, UK

M. Badrul Haider Institute of Business and Accounting, Kwansei Gakuin University, Kobe, Japan

Dorli Harms Centre for Sustainability Management (CSM), Leuphana University, Lüneburg, Germany

Christian Herzig Nottingham Business School, Nottingham Trent University, Nottingham, UK

Jacob Hörisch Centre for Sustainability Management (CSM), Leuphana University, Lüneburg, Germany

Sandor Kerekes Department for Environmental Economics, Corvinus University of Budapest, Budapest, Hungary

Jong-Dae Kim College of Business Administration, Inha University, Incheon, Republic of Korea

Hirotsugu Kitada Faculty of Business Administration, Hosei University, Tokyo, Japan

Katsuhiko Kokubu Graduate School of Business Administration, Kobe University, Kobe, Japan

Ki-Hoon Lee Griffith Business School, Griffith University, Brisbane, Australia

José M. Moneva University of Zaragoza, Zaragoza, Spain

John Morelli Department of Civil Engineering Technology, Environmental Management & Safety, Rochester Institute of Technology, Rochester, NY, USA

Eduardo Ortas University of Zaragoza, Zaragoza, Spain

Guillaume Robin IUT – Department of Marketing Techniques, University Paris-Descartes, Paris, France

Stefan Schaltegger Centre for Sustainability Management (CSM), Leuphana University, Lüneburg, Germany

Jörg E.U. Schmidt School of Business, Institute of Management, University of Applied Sciences and Arts Northwestern Switzerland, Windisch, Switzerland

Andrea Tabi Sustainability Indicators Research Centre, Corvinus University of Budapest, Budapest, Hungary

Sarah Elena Windolph Centre for Sustainability Management (CSM), Leuphana University, Lüneburg, Germany

Abbreviations

€	Euros
AA	AccountAbility
ABIS	The Academy of Business in Society
AFEP	French Association of Large Companies
B2B	Business to Business
BAUM e.V.	German Association of Environmental Management
BCG	Boston Consulting Group
BDA	Confederation of German Employers' Associations
BDI	Federation of German Industries
BRW	Business Review Weekly
BS	British Standard
CAC	French stock-market index
CEO	Chief executive officer
CS	Corporate sustainability
CSB	Corporate Sustainability Barometer
CSM	Centre for Sustainability Management
CSR	Corporate social responsibility
DJSI	Dow Jones Sustainability Index
e.V.	German for registered association
EBEB	European Business Environmental Barometer
eco	Ecological
EFQM	European Foundation for Quality Management
EMAS	Eco Management and Audit Scheme
EMS	Environmental management system
Envtl	Environmental
EU	European Union
FEB	Belgian Enterprises Federation
Fig	Figure
GDP	Gross domestic product
GNP	Gross national product
GRI	Global Reporting Initiative

HR	Human resources
ICSB	International Corporate Sustainability Barometer
IMF	International Monetary Fund
Intl	International
ISO	International Organization for Standardization
KSH	Hungarian Statistical Office
LCA	Life cycle assessment
MEDEF	Movement of the Enterprises of France
Mgmt	Management
MIT	Massachusetts Institute of Technology
MNC	Multinational corporation
Natl	National
NGER	National Greenhouse and Energy Reporting
NGO	Non-governmental organisation
NRE	New Economic Regulations
OECD	Organisation for Economic Co-operation and Development
OHSAS	Occupational Health and Safety Assessment Series
Org	Organisation
PCF	Product carbon footprint
PR	Public relations
PwC	PricewaterhouseCoopers
QMS	Quality management system
R&D	Research and development
SA	Social accountability
SBF	French stock-market index
S-EFQM	Sustainability-EFQM
SMEs	Small and medium-sized enterprises
SMS	Social management system
SPSS	Statistical Package for the Social Sciences
Sust	Sustainability
UK	United Kingdom
UN	United Nations
UNEP	United Nations Environment Programme
UNESCO	United Nations Educational, Scientific and Cultural Organization
USA	United States of America
WBCSD	World Business Council for Sustainable Development

List of Figures

List of Tables

Part I
Approach and Overall Results

Chapter 1
International Corporate Sustainability Barometer: Introduction and Structure

Dorli Harms, Jacob Hörisch, Stefan Schaltegger, and Sarah Elena Windolph

Abstract This chapter briefly summarises the development of the debate on corporate sustainability in academia and practice since the 1960s. Building on this synopsis, the edited volume is positioned within the contemporary discussions on sustainability management.

Furthermore, the chapter introduces the conceptual structure of the book chapters, the 'Triple I' approach, which is used to present the findings from each country in a consistent manner. The 'Triple I' approach constitutes a comprehensive scheme for analysing corporate sustainability, as it focuses on the companies' intention to pursue sustainability management (i.e. motivation; issues), the integration of sustainability in the organisation (i.e. connecting sustainability to the core business; involving corporate functions; using drivers of business cases for sustainability) and the actual implementation of sustainability management measures (i.e. stakeholder management; sustainability management tools and standards; measurements).

Likewise, the chapter sets up a framework for comparing the international results, which allows positioning each country within the international context.

Lastly, the reader is provided with an overview of the edited volumes chapters and their key contributions.

D. Harms (✉) • J. Hörisch • S. Schaltegger • S.E. Windolph
Centre for Sustainability Management (CSM), Leuphana University, Lüneburg, Germany
e-mail: dharms@uni.leuphana.de; hoerisch@uni.leuphana.de; schaltegger@uni.leuphana.de; windolph@uni.leuphana.de

S. Schaltegger et al. (eds.), *Corporate Sustainability in International Comparison*, Eco-Efficiency in Industry and Science 31, DOI 10.1007/978-3-319-06227-3_1,
© Springer International Publishing Switzerland 2014

1.1 Corporate Sustainability Management: An Emerging Field in Business and Academia

In the 1960s and 1970s the predominant goal of business was profit maximisation. Since the late 1990s, however, companies have been increasingly challenged to consider criteria other than economic profitability (Shrivastava 1995; Elkington 1997). Today, companies are expected to make profits and simultaneously act in environmentally and socially responsible ways (Carroll and Buchholtz 2006; Epstein 2008; Munda 2008; Carroll and Shabana 2010).

This edited volume investigates how companies deal with these altered expectations and with these new challenges by depicting the current state and progress of sustainability management in corporate practice. It focuses on three areas of corporate sustainability: its intention, its integration and its implementation. The book builds on the *International Corporate Sustainability Barometer* (ICSB), a survey carried out among *large companies in 11 countries from 4 continents* (see also Schaltegger et al. 2013).

This empirical survey provides valuable insights into what corporate sustainability management comprises on an international scale. The comparative analysis of the status quo of sustainability management practices in various countries allows the identification of global patterns, i.e. cross-national similarities and differences.

In addition to the increasing academic interest in corporate sustainability, in corporate practice several sustainability issues such as energy and water consumption or consumer protection have gained importance over the last decades (e.g. Haanaes et al. 2012; Kiron et al. 2013). Corporate sustainability management, on the one hand, means that companies are challenged to integrate social and environmental issues into their core business. On the other, it requires companies to contribute to a sustainable development of society and the economy (Shrivastava and Hart 1995; Schaltegger et al. 2002; Bansal 2005; Schaltegger and Burritt 2005; Starik and Kanashiro 2013).

In a globalised and networked world economy, it could be expected that managers address sustainability challenges similarly, regardless of country-specific characteristics. Alternatively, it could be argued that the degree of engagement for sustainable development varies according to country-specific historical and cultural backgrounds as well as in the resulting environmental, social and economic conditions. To assess these opposing views, this edited volume investigates the differences as well as the similarities between corporate sustainability management practices in 11 economically developed countries.

1.2 Research Framework: The 'Triple I' Approach

The ICSB has developed a 'Triple I' approach to analyse the three main research areas (see Fig. 1.1) of the practice of corporate sustainability management on an international level.

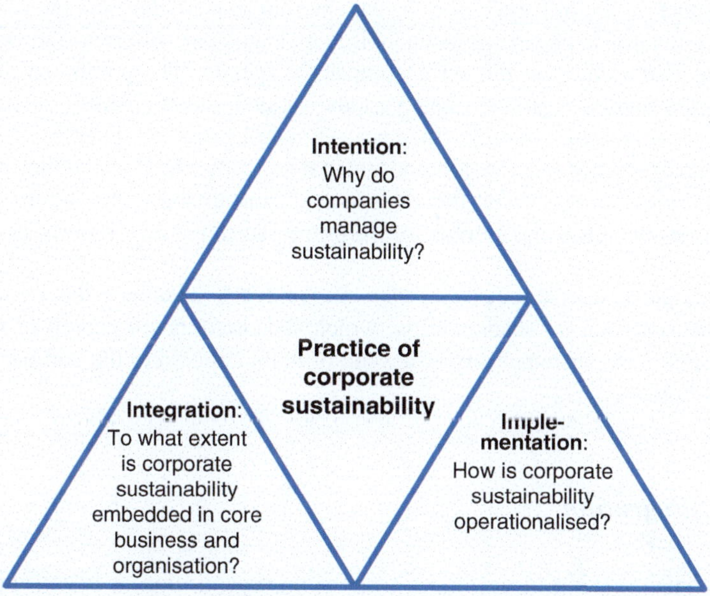

Fig. 1.1 'Triple I' approach of the ICSB

The 'Triple I' approach offers a structured way to comprehensively analyse the various aspects of corporate sustainability. It covers (1) *intention,* the motivations and objectives of corporate sustainability; (2) *integration,* the company's embedding of sustainability into its business and organisation; as well as (3) *implementation,* the dissemination of sustainability measures and tools and the measurement of sustainability impacts within the company.

1.2.1 Intention

There are a number of reasons why corporate actors are motivated to engage in sustainability management. The type of motivation may lead to different activities and often a different degree of engagement. Receiving corporate legitimacy and achieving market success are regarded as complementary corporate motivations to engage with sustainability (Bansal and Roth 2000; Epstein 2008). On the one hand, striving for legitimacy can be considered a reaction to sustainability-related regulations and pressures from societal stakeholders, so-called 'push factors' (Black and Härtel 2004). Market success, on the other, is a motivation for corporate sustainability if consumers or investors offer sustainability-related incentives through purchasing and investing in sustainability-oriented products and services, so-called 'pull factors' (see for example Dunphy et al. 2007; Babiak and Trendafilova 2011; Ditlev-Simonsen and Midttun 2011; Windolph et al. 2013).

Depending on the influence of different external stakeholders and the company-internal motivation to engage in sustainability management, different environmental and social issues may be put on a company's agenda. The continued attention given by the media, regulators and various societal stakeholders to such topics as resource availability, energy efficiency, climate change or diversity illustrates how sustainability has become a broad societal concern which constitutes a global megatrend. One indicator for the relevance of sustainability for companies is, for example, the growing number of corporate sustainability reports (e.g. GRI 2012).

In sum, the research focus *intention* of the ICSB examines the reasons for companies to engage in sustainability, which stakeholders are perceived to exert influence and what sustainability issues are of main concern to the companies and their stakeholders.

1.2.2 Integration

For companies seeking to develop a sustainable organisation, it is not enough to consider environmental and social engagement solely as a cost-intensive optional 'add-on' to their business. Although philanthropic engagement has undoubtedly led to many beneficial projects for society, such activity is not sufficient to achieve sustainable development. As long as a company's core business activities are not themselves sustainable, philanthropic societal engagement will remain superficial and have a kind of 'repair' effect (e.g. Weber 2008). The integration of sustainability into the core business and all organisational units (departments) is therefore crucial for corporate sustainable development. A genuine transformation towards corporate sustainability requires a company to connect its engagement to its economic success and so create business cases for sustainability (see e.g. Schaltegger and Lüdeke-Freund 2012). Innovative answers to the questions how value is created and whether this is done sustainably have a fundamental impact on the economy and society. Creating economic value in a sustainable manner may make the redistribution of (unsustainably) created value redundant.

From a business perspective, moreover, it can be argued that companies which pay little or no attention to the linkages between business, the environment and society are "missing many new sustainable development [. . .] opportunities that may prevent the threat of an inevitable collapse of society" (Moore and Manring 2009: 276). By gaining competitive advantage through sustainability engagement, companies create so-called business cases for sustainability. These are "characterized by creating economic success *through* (and not only along with) a certain environmental or social activity" (Schaltegger and Lüdeke-Freund 2012: II).

Business cases can be created either through externally-oriented measures (such as increasing corporate reputation) or through internally-oriented measures which contribute to economic benefits (such as increasing efficiency). The existing literature shows there are seven potential drivers of business cases for sustainability

(WBCSD 2002; Epstein and Roy 2003; Steger 2004; Schaltegger and Wagner 2006; Schaltegger 2011; Schaltegger et al. 2012).

To incorporate sustainability as an integral part of a company's core business, the existing literature furthermore argues that all organisational units should be involved in corporate sustainability (Hoffman 2001; Dunphy et al. 2007; Epstein 2008; Lauring and Thomsen 2009; Windolph et al. 2013; Schaltegger et al. 2014). A sustainable redesign of value-creation activities and the business model requires the motivation of all corporate actors and coordination between product design, production, logistics, marketing, etc. throughout the whole organisation. The integration of corporate sustainability into routine business operations can and often has to be supported by managers and employees in various corporate functions such as purchasing, manufacturing, research & development (R&D), sales and marketing. Moreover, cross-functional coordination seems to be essential to prevent sustainability issues from being superficially addressed and from being shifted back and forth between functional areas (Shrivastava and Hart 1995; Dunphy et al. 2007; Martin et al. 2007; Epstein 2008). In contrast, cross-functional collaboration may contribute to creating more comprehensive sustainability solutions (Schaltegger et al. 2014).

In sum, the *integration* section of the ICSB addresses the linkage of sustainability with the core business, the integration into organisational units and the drivers of business cases for sustainability.

1.2.3 Implementation

The ICSB addresses the actual implementation of corporate sustainability by surveying the management of stakeholder relationships, the awareness and application of specific sustainability management tools and standards as well as the measurement of corporate sustainability performance and impacts.

Stakeholder relationship management is a core issue for sustainable development and can be of strategic relevance for a company. Stakeholders are defined as groups or individuals who are able to affect the achievement of corporate goals or, vice versa, are themselves affected by a company (Freeman 1984, 2010). The participation of societal groups, and thus their involvement as stakeholders, is a core goal of sustainable development and has already been addressed extensively in the Rio Declaration (United Nations 1992). The management of stakeholder relationships is not only of importance for governments and political organisations but also for companies. Stakeholder relationship management may furthermore be beneficial for companies, since stakeholders enable corporations to gain an external perspective on their sustainability performance. The feedback provided by stakeholders may help companies to detect societal trends and anticipate or react quickly to external change. If such changes are addressed in time, companies can profit by providing customers with sustainability-oriented innovations (Ruppel and Harrington 2000; Harting et al. 2006; Hansen et al. 2009).

Depending on the degree companies want to involve their stakeholders in sustainability efforts, they can manage their stakeholder relationships in different ways. Stakeholder relationship management can range from more passive forms such as observing and informing stakeholders to more participative forms such as empowering them or delegating decision-making authority (modified from Krick et al. 2005).

To enable companies to implement these forms of stakeholder relationships and to operationalise sustainability strategies, various management tools such as stakeholder dialogues, community advisory panels or corporate volunteering have been developed. Besides these stakeholder-oriented tools, there are many other sustainability management tools which can be applied in different organisational units and which serve different purposes. Sustainability or environmental reports, for instance, assist companies in communicating their sustainability challenges, activities and performance. Sustainable design supports the production as well as the R&D units in developing sustainability-oriented innovations. Environmental or corporate social responsibility (CSR) departments can use environmental management systems to better monitor and manage corporate sustainability activities. To investigate the diffusion of these tools in corporate practice, the ICSB surveys corporate awareness and application of them in detail. The ICSB also examines the dissemination of 12 sustainability-related guidelines and standards (such as ISO 14001, GRI) to analyse their role in sustainability management practice.

Finally, the *implementation* of corporate sustainability is reflected in the measurement of the sustainability effects of the projects and activities undertaken and thus of the (economic) success of corporate sustainability. Therefore, the third aspect of implementation surveyed in the ICSB is the measurement of the environmental, social and economic impacts of a company's sustainability management practices. The measurement of a company's influence on environmental and social issues is necessary if the company is to monitor improvements and setbacks. The measurement of business case drivers, moreover, is needed to identify and realise the economic gains associated with sustainability, which in turn helps to anchor sustainability in core business activities (Schaltegger and Wagner 2006; Epstein 2008).

1.3 Analytical Framework

The findings of the ICSB are presented using the 'Triple I' framework. It is applied to the chapter on the international aggregate results, the country-specific chapters, and on the comparative analysis. The idea behind the 'Triple I' approach is to take a comprehensive view of the practice of corporate sustainability management and to identify particular national and international characteristics that affect sustainability.

The framework, first of all, aims at contrasting society-oriented and market-oriented drivers and the *intentions* behind corporate efforts to manage sustainability. Aspects such as relevant stakeholders or expected impacts from implementing corporate sustainability can be categorised as society-oriented (e.g. NGOs, enhancing and safeguarding corporate reputation) or market-oriented (e.g. customers, revenue increase). This categorisation helps to describe patterns of what drives companies to engage in sustainability management.

Second, the framework examines how companies *integrate* social and environmental issues into their business activities. Here, companies can be clustered into those that are defensive and those that are proactive in their approach to corporate sustainability management. A company's commitment to sustainability is analysed in terms of whether it is connected to the company's core business. The analysis also provides a quantitative and qualitative assessment of which functional units are involved in corporate sustainability.

Finally, the aspect of the *implementation* of corporate sustainability assesses how sustainability management is operationalised by the companies. Here, both perspectives – society vs. market orientation and defensive vs. proactive measures – are combined and extended by evaluating what companies actually do, what kind of sustainability management tools they know of and apply and if they measure the impacts of their sustainability management.

For the interpretation of the results further factors, such as the sectoral structure, of the countries are analysed and the company size in terms of annual revenue and the number of employees is also taken into account.

To compare the country-specific findings, the results are examined using three measures. First, the *spread* of the national averages reveals the extent of differences between different countries with regard to specific aspects of sustainability management. Second, common global *patterns* are identified which are shared by most countries. Third, the general intensity of corporate practices in different countries is compared to identify which countries show outstanding overall results.

1.4 Overview of Chapters and Contributions

This edited volume on the practice of corporate sustainability management describes and compares the status quo of sustainability management and CSR on an international level by adopting the 'Triple I' approach.

In Chap. 2 Jacob Hörisch outlines the *Purpose and Approach* of the Barometer survey by identifying the research gap and describing the methodology. In Chap. 3 Jacob Hörisch and Sarah Elena Windolph give an *Overview of the Aggregate Results of the International Corporate Sustainability Barometer.* These average results and the data of all 11 countries participating in the ICSB serve as a reference point and benchmark for the following country-specific chapters of the 2012 ICSB survey.

Part II contains the *Country-Specific Findings* that reveal characteristics and particularities of the practice of sustainability management in the countries surveyed from Europe, Asia, Australia and North America. Country-specific insights from distinguished researchers involved in the ICSB project allow for comparison and discussion of the national and international findings. In Chap. 4 Roger Burritt and Amanda Carter analyse the ICSB results from *Australia*. In Chap. 5, Nathalie Crutzen presents the findings for *Belgium*. Chapter 6, by Amel Ben Rhouma, Claude Francoeur and Guillaume Robin, provides the specific findings for *France*. In Chap. 7 Jacob Hörisch, Sarah Elena Windolph and Stefan Schaltegger describe the *German* results. In Chap. 8 Maria Csutora, Andrea Tabi and Sandor Kerekes add further European insights into the practice of sustainability management in *Hungary*. In Chap. 9, Katsuhiko Kokubu, Hirotsugu Kitada and Mohammad Badrul Haider present the results of the ICSB in *Japan*. In Chap. 10, Jong-Dae Kim and Ki-Hoon Lee describe the findings from the *Korean* survey. In Chap. 11 José M. Moneva, Eduardo Ortas and Igor Álvarez analyse country-specific results by exploring sustainability in *Spanish* companies. In Chap. 12, Jörg E.U. Schmidt and Claus-Heinrich Daub examine the state of the art and progress of corporate sustainability in *Switzerland*. In Chap. 13, Biswaraj Ghosh and Christian Herzig discuss further international findings from the *UK*. As the last of 11 country-specific chapters, in Chap. 14 John Morelli and Dorli Harms analyse the case of corporate sustainability in the *United States of America*.

Part III of this edited volume reveals *Patterns and Conclusions*. In Chap. 15 Stefan Schaltegger and Dorli Harms discuss both similarities and differences for the countries investigated in this survey and draw conclusions for management and future research.

References

Babiak K, Trendafilova S (2011) CSR and environmental responsibility: motives and pressures to adopt green management practices. Corp Soc Responsib Environ Manag 18(1):11–24

Bansal P (2005) Evolving sustainably: a longitudinal study of corporate sustainable development. Strateg Manag J 26(3):197–218

Bansal P, Roth K (2000) Why companies go green: a model of ecological responsiveness. Acad Manag J 43(4):717–736

Black LD, Härtel CEJ (2004) The five capabilities of socially responsible companies. J Public Aff 4(2):125–144

Carroll AB, Buchholtz AK (2006) Business & society: ethics and stakeholder management, 6th edn. Mason, South-Western

Carroll AB, Shabana KM (2010) The business case for corporate social responsibility. A review of concepts, research and practice. Int J Manag Rev 12(1):85–105

Ditlev-Simonsen CD, Midttun A (2011) What motivates managers to pursue corporate responsibility? A survey among key stakeholders. Corp Soc Responsib Environ Manag 18(1):25–38

Dunphy D, Griffiths A, Benn S (2007) Organizational change for corporate sustainability. A guide for leaders and change agents of the future, 2nd edn. Routledge, London

Elkington J (1997) Cannibals with forks: the triple bottom line of 21st century business. Capstone, Oxford

Epstein MJ (2008) Making sustainability work. Best practices in managing and measuring corporate social, environmental, and economic impacts. Greenleaf, Sheffield

Epstein MJ, Roy M-J (2003) Making the business case for sustainability: linking social and environmental actions to financial performance. J Corp Citizen 9:79–96

Freeman RE (1984) Strategic management: a stakeholder approach. Pitman, Boston

Freeman RE (2010) Stakeholder theory: the state of the art. Cambridge University Press, Cambridge/New York

Global Reporting Initiative (GRI) (2012) G3 guidelines. https://www.globalreporting.org/reporting/reporting-framework-overview/Pages/default.aspx. Accessed 7 Jul 2013

Haanaes K, Reeves M, von Streng Velken I, Audretsch M, Kiron D, Kruschwitz N (2012) Sustainability nears a tipping point, MIT Sloan Management Review. Massachusetts Institute of Technology/Boston Consulting Group, North Hollywood

Hansen EG, Grosse-Dunker F, Reichwald R (2009) Sustainability innovation cube – a framework to evaluate sustainability-oriented innovations. Int J Innov Manag 13(4):683–713

Harting T, Harmeling SS, Venkataraman S (2006) Innovative stakeholder relations. When ethics pays (and when it doesn't). Bus Ethics Q 16(1):43–68

Hoffman AJ (2001) Linking organizational and field-level analyses: the diffusion of corporate environmental practice. Organ Environ 14(2):133–156

Kiron D, Kruschwitz N, Haanaes K, Reeves M, Goh E (2013) The innovation bottom line. How companies that see sustainability as both a necessity and an opportunity, and change their business models in response, are finding success. Massachusetts Institute of Technology/Boston Consulting Group, Boston

Krick T, Forstater M, Monaghan P, Sillanpää M (2005) Stakeholder engagement manual, vol 2, The practitioner's handbook on stakeholder engagement. AccountAbility, London

Lauring J, Thomsen C (2009) Collective ideals and practices in sustainable development: managing corporate identity. Corp Soc Responsib Environ Manag 16(1):38–47

Martin A, Benn S, Dunphy D (2007) Towards a model of governance for sustainability. In: Benn S, Dunphy D (eds) Corporate governance and sustainability: challenges for theory and practice. Routledge, London, pp 94–121

Moore SB, Manring SL (2009) Strategy development in small and medium sized enterprises for sustainability and increased value creation. J Clean Prod 17(2):276–282

Munda G (2008) Social multi-criteria evaluation for a sustainable economy. Springer, Berlin/Heidelberg

Ruppel CP, Harrington SJ (2000) The relationship of communication, ethical work climate, and trust to commitment and innovation. J Bus Ethics 25(4):313–328

Schaltegger S (2011) Sustainability as a driver for corporate economic success. Consequences for the development of sustainability management control. Soc Econ 33(1):15–28

Schaltegger S, Burritt R (2005) Corporate sustainability. In: Folmer H, Tietenberg T (eds) The international yearbook of environmental and resource economics. Edward Elgar, Cheltenham, pp 185–232

Schaltegger S, Lüdeke-Freund F (2012) The "business case for sustainability" concept: a short introduction. Centre for Sustainability Management, Lüneburg

Schaltegger S, Herzig C, Kleiber C, Müller J (2002) Sustainability management in business enterprises, concepts and instruments for sustainable organisation development. The Federal Ministry for the Environment, Nature Conservation and Nuclear Safety (BMU)/Federation of German Industries (BDI), Bonn/Berlin

Schaltegger S, Lüdeke-Freund F, Hansen EG (2012) Business cases for sustainability: the role of business model innovation for corporate sustainability. Int J Innov Sustain Dev 6(2):95–119

Schaltegger S, Harms D, Hörisch J, Windolph SE, Burritt R, Carter A, Truran S, Crutzen N, Ben-Rhouma A, Csutora M, Tabi A, Kokubu K, Kitada H, Haider MB, Kim J-D, Lee K-H, Moneva JM, Ortas E, Álvarez-Etxeberria I, Daub C-H, Schmidt J, Herzig C, Morelli J (2013) International corporate sustainability barometer: a comparative study of 11 countries. Centre for Sustainability Management, Lüneburg

Schaltegger S, Harms D, Windolph SE, Hörisch J (2014) Involving corporate functions: who contributes to sustainable development? Sustainability 6(5):3064–3085

Schaltegger S, Wagner M (2006) Managing the business case for sustainability. The integration of social, environmental and economic performance. Greenleaf, Sheffield

Shrivastava P (1995) The role of corporations in achieving ecological sustainability. Acad Manag Rev 20(4):936–960

Shrivastava P, Hart S (1995) Creating sustainable corporations. Bus Strateg Environ 4(3):154–165

Starik M, Kanashiro P (2013) Toward a theory of sustainability management: uncovering and integrating the nearly obvious. Organ Environ 26(1):7–30

Steger U (ed) (2004) The business of sustainability: building industry cases for corporate sustainability. Palgrave Macmillan, Houndmills

United Nations (1992) Rio declaration on environment and development 1992. United Nations Environment Programme (UNEP), New York

Weber M (2008) The business case for corporate social responsibility: a company-level measurement approach for CSR. J Environ Manag 26(4):247–261

Windolph SE, Harms D, Schaltegger S (2013) Motivations for corporate sustainability management: contrasting survey results and implementation. Corp Soc Responsib Environ Manag (online first)

World Business Council for Sustainable Development (WBCSD) (2002) The business case for sustainable development: making a difference towards the Earth Summit 2002 and beyond. Corp Environ Strateg 9(3):226–235

Chapter 2
International Corporate Sustainability Barometer: Purpose and Approach

Jacob Hörisch

Abstract In a first step, this chapter provides a brief review of the existing literature on the international state of the art of sustainability management. The review reveals that while a large body of studies already exists, most publications deal with single nations or bilateral comparisons. Among the truly international studies, a lack of comparative quantitative inquiries on a country-specific level is identified.

In a second step, the edited volume's methodological approach to address this research gap is outlined. Unlike prior studies on sustainability management, the ICSB builds on a comprehensive survey among sustainability managers in large companies and is not restricted to single aspects of sustainability management. This chapter describes the process of data collection and provides the reader with the most important information on the resulting sample which includes 468 companies of 11 economically developed countries from 4 distinct world regions.

2.1 Current State of Research: Identifying the Research Gap

The growing importance of sustainability issues for society, politics and business (see Chap. 1) not only motivates companies to establish sustainability management practices, but it has also led to an increasing number of academic publications on sustainability management. The German 'Corporate Sustainability Barometer' (Schaltegger et al. 2010) conducted a literature review and identified a growing number of sustainability management articles in sustainability-related journals as well as in conventional management journals. Sustainability management has become an integral part of mainstream business studies.

J. Hörisch (✉)
Centre for Sustainability Management (CSM), Leuphana University, Lüneburg, Germany
e-mail: hoerisch@uni.leuphana.de

S. Schaltegger et al. (eds.), *Corporate Sustainability in International Comparison*, Eco-Efficiency in Industry and Science 31, DOI 10.1007/978-3-319-06227-3_2,
© Springer International Publishing Switzerland 2014

Numerous studies have been conducted since the late 1990s to describe the status quo of sustainability management and corporate social responsibility (CSR) practices. National analyses exist for most European countries (e.g. Harkai and Pataki 2001; Pacheco and Wehrmeyer 2001; Wagner and Schaltegger 2002; Bertelsmann Foundation 2005; Antal and Sobczak 2007; Delbard 2008; Schaltegger et al. 2010) as well as for numerous countries in North and Latin America (e.g. Cecil 2008; United Nations 2012) and in the Asia-Pacific region (e.g. Frost et al. 2005) to name just a small selection of studies on the countries involved in the International Corporate Sustainability Barometer (see also Schaltegger et al. 2013). Besides country-specific studies, empirical research has been primarily concentrated on bilateral comparisons (James et al. 1997; Bansal and Roth 2000; Wehrmeyer et al. 2002). However, these studies frequently concentrate on single aspects of sustainability management such as sustainability reporting (Frost et al. 2005; Cecil 2008) or corporate environmental strategies (Wehrmeyer et al. 2002).

An important step towards multi-country comparative analysis in the field of sustainability management was the European Business Environmental Barometer (EBEB). This quantitative survey conducted in 11 European countries in 1997/98 covered a wide range of environmental management topics. It resulted in various publications (e.g. Baumast 2000; Baumast and Dyllick 2001; Kestemont and Ytterhus 2001) on national differences in management practices as well as on the role of contextual factors such as national market forces, legislation or culture. However, the results date back to the late 1990s and do not cover any non-European countries. The EBEB's focus on environmental issues furthermore means that it does not deal with a number of central aspects of sustainability management, such as balancing and simultaneously managing social, environmental and financial aspects.

Until today, there is a lack of truly international academic studies on sustainability management. Most publications concentrate on a relatively small number of countries and on specific aspects (such as the application of ISO 14001 as an environmental management standard) related to sustainability management (e.g. Wehrmeyer et al. 2002; Burritt et al. 2003), use secondary data, or they collect comparatively general qualitative, primarily practitioner-oriented data. These studies do not allow then for comprehensive in-depth quantitative comparisons. For instance, although the journal articles by Neumayer and Perkins (2004) as well as Delmas and Montiel (2008) analyse companies in more than 100 different countries, their analyses are limited to the diffusion of voluntary management standards. Delmas and Montiel (2008) restrict their sample to companies in the chemical industry. Similarly, although Forbes and McIntosh (2011) use existing indicators to compare CSR practices in 16 countries, their analysis focuses on the Asia-Pacific region, excluding companies from other economically advanced regions.

One of the first international studies of corporate sustainability management practices was documented in the UN Global Compact Annual Review of 2007 (UN Global Compact 2007). In 2010, the UN Global Compact, together with the consulting firm Accenture, extended its study and surveyed 766 CEOs (Lacy et al. 2010).

However, with their focus on CEOs both studies primarily survey managerial attitudes and expectations towards sustainability rather than depicting the current practice of corporate sustainability management. Furthermore, the study almost exclusively collects qualitative information and does not aim at nation-specific comparisons of implementation activities. Instead it gives a practitioner-oriented overview of attitudes towards sustainability and adds some comparisons on a continent-specific level.

Similarly, the annual reports published by the MIT Sloan Management Review and the consulting firm The Boston Consulting Group (BCG) provide very valuable insights by surveying more than 1,500 corporate leaders from all over the world in their annual surveys (Berns et al. 2009; Haanaes et al. 2011, 2012; Kiron et al. 2013). Again, due to the focus on the perceptions of corporate leaders the emphasis is more on attitudes than on measures actually taken, and the questionnaire concentrates on qualitative information. As in the study by Accenture and UN Global compact (Lacy et al. 2010), the results are not displayed on a country-specific basis, but instead global trends are identified and global regions are compared.

Other publications such as the UN Global Compact International Yearbooks (e.g. UN Global Compact 2012) illustrate the practice of sustainability management in selected companies all over the world. However, these publications tend to focus on promoting best practice examples rather than on identifying and analysing the actual status quo of corporate sustainability practice. Idowu and Leal Filho (2009), taking a different approach, aim at systematically documenting the current state of CSR in 19 countries from Europe, East Asia, Middle East/Africa and the Americas. However, these findings are not based on primary survey data but provide what are often conceptual summaries of the national CSR frameworks drawing on anecdotal examples from single companies. Due to the focus on CSR instead of sustainability management, the authors largely neglect the application of a broad range of sustainability management tools and how sustainability can be integrated into the company's core economic activities.

This brief review of the literature reveals that while a large body of studies already exists, most publications deal with single nations or bilateral comparisons. Among the truly international studies, there is a lack of in-depth academic investigations on a comparative but country-specific basis. Most attempts focus on qualitative information, draw global trends and are primarily oriented to practitioners.

More academically-oriented, quantitative publications usually focus on single aspects of sustainability management. In most analyses, the perceptions and attitudes of CEOs and corporate leaders are surveyed. However, sustainability managers or other operational middle managers explicitly authorised to deal with sustainability may be better able to provide a valid insight not only into attitudes but also into the details of actual sustainability management practices.

This review shows that a research gap exists for a large-scale, comparative international study on sustainability management which surveys and compares corporate practices between different countries. This research gap is addressed by the International Corporate Sustainability Barometer (ICSB).

2.2 Methodology: Addressing the Research Gap

In contrast to the international studies described above, the ICSB collected data on a country-specific basis. To be able to examine the effects of different contextual factors (e.g. markets, legislation and culture) the 'most-different system design' has been adopted from political sciences (Anckar 2008). Companies from a number of economically developed countries differing in culture, economic size, market structures and traditions were investigated with regard to their sustainability management practices within a single survey:

- For continental Europe, French and German companies were examined, to provide an insight into large Western European economies, whereas Belgian and Swiss companies represent the group of smaller European economies. Central Europe is captured by Hungary and for Southern Europe Spain was examined.
- The United Kingdom was surveyed as a traditionally strongly free-market oriented economy. Likewise, the USA, the largest North American economy and another free-market oriented economy strongly influenced by Anglo-Saxon culture, was also included in the analysis.
- Within the Asia-Pacific region, Japan and the Republic of Korea were surveyed, representing the largest capitalistic industrial economies in East Asia. Last but not least, Australia was included in the survey, as an Asia-Pacific economy distinct from the Asian countries.

This international project was coordinated by the Centre for Sustainability Management (CSM) at Leuphana University Lüneburg, Germany. In each country, a national academic institution organised the country-specific surveys (see Table 2.1).

Unlike other empirical sustainability management studies, the questionnaire used for the ICSB was not restricted to a single aspect of sustainability management or a single environmental or social topic. It aimed at covering a broad range of corporate sustainability issues, with a special focus on a company's *intention* to pursue sustainability management, the *integration* of sustainability into its business activities and its actual *implementation*. In contrast to more practitioner-oriented, qualitative studies, this research instrument was largely based on quantitative questions. These served to describe and analyse quantitative differences for the surveyed aspects instead of comparing percentages of dichotomous answers.

The questionnaire was developed by the CSM and made use of the experience of prior surveys conducted by the CSM on sustainability management in Germany in 2002, 2006 and 2010. Where necessary, the national academic institutions translated the questionnaire into the country's main language. Back translations were organised to ensure that in each country the same questions were asked in exactly the same manner. This procedure served to enable multi-country comparisons. Before the survey started, extensive pre-testing was conducted to validate the questionnaire. To reduce the effort for participating companies, an online-questionnaire was used. All national surveys were carried out between February and August 2012.

Table 2.1 Participating research institutions, responses and response rates in the surveyed countries

Country	Institution	Responses (quantity/ response rate [%])
Australia	University of South Australia; Centre of Accounting, Governance and Sustainability	48/26.2
Belgium	University of Liege; HEC Management School	22/15.9
France	University Paris Descartes; Département Gestion des Entreprises et des Administrations	20/21.5
Germany	Leuphana University Lüneburg; Centre for Sustainability Management (CSM)	152/39.7
Hungary	Corvinus University of Budapest; Department of Environmental Economics and Technology	28/32.9
Japan	Kobe University; Graduate School of Business Administration	48/16.0
Korea	Inha University; Sustainability Management Research Institute	32/15.5
Spain	University of Zaragoza; Faculty of Economics and Business Administration	23/26.1
	University of Basque Country; Faculty of Economics and Business Studies	
Switzerland	University of Applied Sciences and Arts Northwestern Switzerland; School of Business; Institute of Management	25/12.2
United Kingdom	Nottingham University Business School; International Centre for Corporate Social Responsibility	36/16.4
USA	Rochester Institute of Technology; Department of Civil Engineering Technology, Environmental Management and Safety	34/19.4
Overall		468/22.5

Before the surveys were conducted, in each country lists of the largest companies from all industries were compiled. In most countries, existing databases (e.g. Fortune 500, Welt Online, BelFirst) were used to identify the largest companies by revenue. To avoid double-counting of responses, subsidiaries which do not manage their sustainability management independently were excluded from the lists. Similarly, all companies which ex post turned out not to fit the sample were excluded from the analysis to ensure that the companies from all countries in the study fulfil the same criteria. The most frequent reason to exclude a company from the database was if it reported annual revenue below €50 million for the prior financial year, which classified it as a small or medium-sized company according to the EU definition (European Commission 2005). Other companies were excluded ex post if their sustainability management activities were run by a parent company.

In total, 2,076 questionnaires were sent out in the 11 participating countries. Altogether, the international surveys yielded 468 valid responses (response rate of 22.5 %). The ICSB data thus meets the requirements Bartlett et al. (2001) set up for performing statistical analysis using quantitative survey data. Additionally, the response rate is within the standard deviation range Baruch and Holtom (2008) identified as appropriate for survey-based scientific articles on organisations published in refereed academic journals. The country-specific differences in the response rate can be explained by different cultures and attitudes towards surveys (see also O'Neill et al. 1995; de Heer 1999).

For the data analysis, IBM SPSS 20 was used, which allows building and comparing subsamples for each national dataset. The main characteristics of the resulting international sample are displayed in the following chapter.

References

Anckar C (2008) On the applicability of the most similar systems design and the most different systems design in comparative research. Int J Soc Res Methodol 11(5):389–401

Antal AB, Sobczak A (2007) Corporate Social Responsibility in France: A Mix of National Traditions and International Influences. Bus Soc 46(1):9–32

Bansal P, Roth K (2000) Why companies go green: a model of ecological responsiveness. Acad Manag Rev 43(4):717–736

Bartlett JE, Kotrlik JW, Higgins CC (2001) Organizational research: determining appropriate sample size in survey research. Inf Technol Learn Perform J 19(1):43–50

Baruch Y, Holtom BC (2008) Survey response rate levels and trends in organizational research. Hum Relat 61(8):1139–1160

Baumast A (2000) Environmental management in Europe. Results of the European Business Environmental Barometer (E.B.E.B). IWOe discussion paper, vol 79. Institute for Economy and the Environment at the University of St. Gallen, St. Gallen

Baumast A, Dyllick T (eds) (2001) Umweltmanagement-Barometer 2001. IWÖ-Diskussionsbeitrag, vol 93. Institut für Wirtschaft und Ökologie an der Universität St Gallen, St. Gallen

Berns M, Townsend A, Khayat Z, Balagopal B, Reeves M, Hopkins MS, Kruschwitz N (2009) The business of sustainability. Findings and insights from the first annual business of sustainability survey and the Global Thought Leaders' research project. MIT Sloan Management Review; The Boston Consulting Group, North Hollywood

Bertelsmann Foundation (2005) Die gesellschaftliche Verantwortung von Unternehmen. Dokumentation der Ergebnisse einer Unternehmensbefragung der Bertelsmann Stiftung. Bertelsmann Foundation, Gütersloh. http://www.bertelsmann-stiftung.de/cps/rde/xbcr/SID-B7B1C456-51B7A407/bst/Unternehmensbefragung_CSR_200705.pdf. Accessed 9 Aug 2013

Burritt R, Schaltegger S, Kokubu K, Wagner M (2003) Environmental management accounting for staff appraisal: evidence from Australia, Germany and Japan. In: Bennett M, Rikhardsson PM, Schaltegger S (eds) Environmental management accounting – purpose and progress. Kluwer, Dordrecht, pp 151–188

Cecil L (2008) Corporate social responsibility reporting in the United States. McNair Scholar Res J 1(1):43–52

de Heer W (1999) International response trends: results of an international survey. J Off Stat 15(2):129–142

Delbard O (2008) CSR legislation in France and the European regulatory paradox: an analysis of EU CSR policy and sustainability reporting practice. Corp Gov 8(4):397–405

Delmas M, Montiel I (2008) The diffusion of voluntary international management standards: responsible care, ISO 9000, and ISO 14001 in the chemical industry. Pol Stud J 36(1):65–93

European Commission (2005) The new SME definition: user guide and model declaration. http://ec.europa.eu/enterprise/policies/sme/files/sme_definition/sme_user_guide_en.pdf. Accessed 2 Apr 2012

Forbes S, McIntosh M (2011) Towards CSR and the sustainable enterprise economy in the Asia Pacific region. Sustain Account Manage Pol J 2(2):194–213

Frost G, Jones S, Loftus J, van der Laan S (2005) A survey of sustainability reporting practices of Australian reporting entities. Aust Account Rev 15(35):89–96

Haanaes K, Arthur D, Balagopal B, Kong MT, Reeves M, Velken I, Hopkins MS, Kruschwitz N (2011) Sustainability: the 'embracers' seize advantage. MIT Sloan Management Review; The Boston Consulting Group, North Hollywood

Haanaes K, Reeves M, von Strang Velken I, Audretsch M, Kiron D, Kruschwitz N (2012) Sustainability nears a tipping point. MIT Sloan Management Review; Boston Consulting Group, North Hollywood

Harkai A, Pataki G (2001) Bericht zum Umweltmanagementbarometer Ungarn. In: Baumast A, Dyllick T (eds) Umweltmanagement-Barometer 2001, IWÖ-Diskussionsbeitrag, vol 93. Institut für Wirtschaft und Ökologie an der Universität St. Gallen, St. Gallen

Idowu SO, Leal Filho W (eds) (2009) Global practices of corporate social responsibility. Springer, Heidelberg

James P, Prehn M, Steger U (1997) Corporate environmental management in Britain and Germany. Anglo-German Foundation for the Study of Industrial Society, London

Kiron D, Kruschwitz N, Haanaes K, Reeves M, Goh E (2013) The innovation bottom line. How companies that see sustainability as both a necessity and an opportunity, and change their business models in response, are finding success. MIT Sloan Management Review; Boston Consulting Group, North Hollywood

Kestemont M, Ytterhus B (2001) International Business Environment Barometer 1997. Final report to the EC. Catholic University of Louvain, Louvain-la-Neuve

Lacy P, Cooper T, Hayward R, Neuberger L (2010) A new era of sustainability. CEO reflections on progress to date, challenges ahead and the impact of the journey toward a sustainable economy. http://www.accenture.com/SiteCollectionDocuments/PDF/Accenture_A_New_Era_of_Sustainability_CEO_Study.pdf. Accessed 8 May 2014

Neumayer E, Perkins R (2004) What explains the uneven take-up of ISO 14001 at the global level? A panel-data analysis. Environ Plan A 36(5):823–839

O'Neill TW, Marsden D, Matthis C, Raspe H, Silman AJ (1995) Survey response rates: National and regional differences in a European multicentre study of vertebral osteoporosis. J Epidemiol Community Health 49(1):87–93

Pacheco C, Wehrmeyer W (2001) Umweltmanagementstrategien in Großbritannien: Erste Ergebnisse einer Fragebogenuntersuchung. In: Baumast A, Dyllick T (eds) Umweltmanagement-Barometer 2001. IWÖ-Diskussionsbeitrag, vol 93. Institut für Wirtschaft und Ökologie an der Universität St. Gallen, St. Gallen

Schaltegger S, Windolph SE, Harms D (2010) Corporate Sustainability Barometer 2010. Centre for Sustainability Management, Lüneburg

Schaltegger S, Harms D, Hörisch J, Windolph SE, Burritt R, Carter A, Truran S, Crutzen N, Ben Rhouma A, Csutora M, Tabi A, Kokubu K, Kitada H, Haider MB, Kim J-D, Lee K-H, Moneva JM, Ortas E, Álvarez-Etxeberria I, Daub C-H, Schmidt J, Herzig C, Morelli J (2013) International Corporate Sustainability Barometer: a comparative study of 11 countries. Centre for Sustainability Management, Lüneburg

UN Global Compact (2007) UN global compact annual review 2007. United Nations Global Compact Office, New York

United Nations (2012) Global compact international yearbook 2012. United Nations Global Compact Office, New York

Wagner M, Schaltegger S (2002) Umweltmanagement in deutschen Unternehmen. Der aktuelle Stand der Praxis. Centre for Sustainability Management, Lüneburg

Wehrmeyer W, Wagner M, Pacheco C, Schaltegger S (2002) Environmental management strategies: Britain and Germany compared. Anglo-German Foundation for the Study of Industrial Society, London

Chapter 3
Overview of the Aggregate Results of the International Corporate Sustainability Barometer

Jacob Hörisch and Sarah Elena Windolph

Abstract Managing sustainable development issues on the corporate level is generally acknowledged to be of growing importance for companies world-wide. Yet, to date little is known about the practice of sustainability management in different countries and how national practices compare to each other. More specifically, do companies in different regions of the world manage sustainability similarly, that is, is corporate sustainability a global trend and is there an "international state of the art" of sustainability management practice, perhaps even reflecting isomorphic tendencies? Or is sustainability management more of a regionally differentiated phenomenon? Do local challenges, cultural backgrounds and economic factors lead to different approaches towards corporate sustainability? This chapter presents the international average and country-specific data from the International Corporate Sustainability Barometer (ICSB). The international average results serve as a reference point showing the state of the art for the following 11 chapters offering details on the specific results for each country in the 2012 ICSB survey.

3.1 Introduction

Various publications emphasise the world-wide relevance of corporate sustainability management (e.g. Berns et al. 2009; Idowu and Leal Filho 2009; Lacy et al. 2010; Haanaes et al. 2011, 2012; Kiron et al. 2013). Little is known, however, about whether there is a common state of the art in corporate sustainability management in industrialised countries, understood as a common set of practices irrespective of national economic or cultural particularities. This chapter describes the average international findings of the International Corporate Sustainability

J. Hörisch (✉) • S.E. Windolph
Centre for Sustainability Management (CSM), Leuphana University, Lüneburg, Germany
e-mail: hoerisch@uni.leuphana.de; windolph@uni.leuphana.de

S. Schaltegger et al. (eds.), *Corporate Sustainability in International Comparison*, Eco-Efficiency in Industry and Science 31, DOI 10.1007/978-3-319-06227-3_3,

© Springer International Publishing Switzerland 2014

Table 3.1 Sample characteristics – revenue (sample size n = 468)

	Frequency	Percentage
>€50–500 million	65	13.9
>€500–1,500 million	80	17.1
>€1,500–2,500 million	62	13.2
>€2,500–5,000 million	68	14.5
>€5,000–50,000 million	147	31.4
>€50,000 million	46	9.8
Total	468	100.0

Table 3.2 Sample characteristics – number of employees (n = 465)

	Frequency	Percentage
≤50	2	0.4
51–250	13	2.8
251–1,000	45	9.6
1,001–10,000	190	40.6
10,001–100,000	180	38.5
>100,000	35	7.5
Not specified	3	0.6
Total	468	100.0

Barometer (ICSB) survey, which are aggregated based on the responses of all participating companies divided by the number of participating companies. These results serve as an initial reference and benchmark to identify potential national similarities and differences in the following country-specific chapters of the 11 countries participating in the survey (see also Schaltegger et al. 2013).

The international sample consists of 468 companies, each ranking among the largest companies in their home countries (i.e. where the company is located). Since the economies included in the analysis differ with regard to size, with annual GDPs ranging from $126 billion in Hungary to $115,684 billion in the USA (World Bank 2012), it is not surprising that huge differences with regard to company size can be observed among the companies included in the international average. The sample is nevertheless representative for the largest companies in each country and shows a balanced composition of company sizes, thus providing a good database for comparing the sustainability management practices of large companies in different countries. While the annual revenues for 13.9 % of the companies are below €500 million, 9.8 % of the companies have revenues of more than €50,000 million (see Table 3.1).

Smaller differences can be observed for the number of employees, since only 0.4 % of the companies surveyed have 50 or less employees (0.4 %) and no more than 7.5 % have more than 100,000 employees (Table 3.2).

The following sections discuss whether differences can also be observed in sustainability management practices or whether similar patterns can be found, e.g. because of common management standards or isomorphic behaviour (see DiMaggio and Powell 1983).

3.2 Analysis

The international aggregate findings are presented using the 'Triple I' approach developed at the Centre for Sustainability Management (CSM), distinguishing the intention, integration, and implementation of sustainability management practices. Each perspective is further differentiated. Whereas this chapter shows the most important findings on the international level, the country-specific chapters discuss differences and possible reasons for the country-specific results. The concluding chapters then discuss some overarching international patterns and comparisons.

3.2.1 Intention

3.2.1.1 Motivation

In his frequently cited work, Edward Freeman (1984) emphasised the crucial importance of stakeholders in the corporate context. Since then, numerous publications have confirmed the influential role of stakeholders in driving corporate sustainability engagement (e.g. Sharma and Henriques 2005; Darnall et al. 2010; Sarkis et al. 2010; Ditlev-Simonsen and Midtun 2011; Lee 2011).

To empirically examine these findings on an international level, a list of 17 different external stakeholders was presented to the companies participating in the ICSB. These companies were asked to rate each stakeholder's impact on the implementation of sustainability in their company on a five-point semantic differential scale.

Overall, the results support the findings of the literature described above, as on average all external stakeholders are found to promote corporate sustainability engagement and none of the stakeholders surveyed is regarded as primarily inhibiting (see Fig. 3.1). Still, there are significant differences among the stakeholder groups. While most stakeholders are rated as neutral (values ranging around 3) by the majority of companies, more than half of the companies rated competitors, investors/owners/shareholders/cooperative members, national authorities/legislators, the community, international authorities, media/public and, most significantly, NGOs/environmental/social organisations as promoting sustainability management. Banks/creditors, insurance companies and trade unions are the least promoting stakeholders on international average.

3.2.1.2 Issues

Apart from contextual factors (e.g. legislation or public awareness), company-specific characteristics (e.g. corporate and business strategies), its core business and its stakeholders have an influence on which sustainability-related issues a company

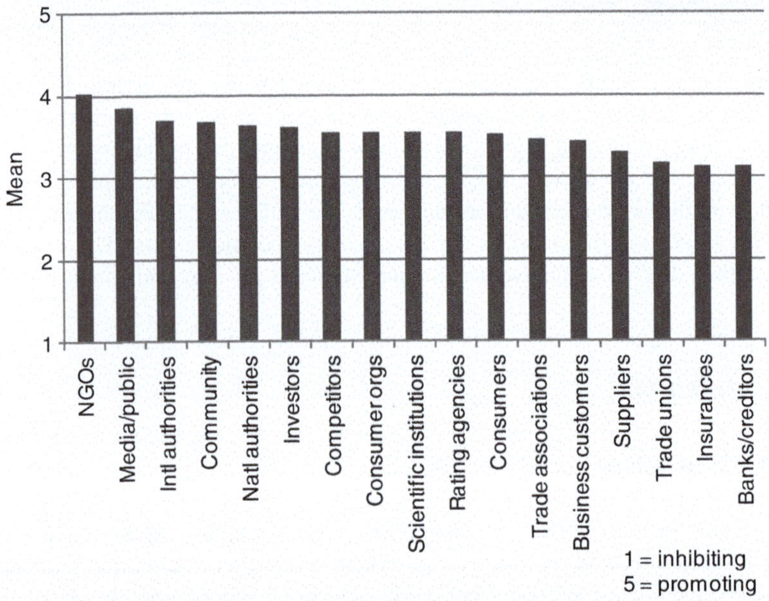

Fig. 3.1 Influence of external stakeholders (n ranging from 393 to 450)
Question: "How would you assess the impact of the following organisational units/stakeholders on the implementation of sustainability in your company?"

tends to manage. A five-point semantic differential scale measures the extent to which companies manage the 13 sustainability issues of interest in this study.

The resulting pattern is twofold (see Fig. 3.2). Some sustainability issues – emissions/wastewater/waste, workplace/employment, training and development, energy consumption, occupational health and safety – seem to be of central importance for nearly all companies globally. They are closely managed in more than half of all companies and the standard deviation is relatively small, indicating that the differences between the companies surveyed are small. In contrast, biodiversity as well as child labour, forced and compulsory labour are less closely managed, and larger inter-company differences exist. The standard deviation is up to two times higher than for the issues more closely managed (e.g. 0.72 for health and safety as opposed to 1.42 for child labour, forced and compulsory labour).

This demonstrates that, on the one hand, a range of environmental and social issues is closely managed, be it due to external pressures relevant worldwide or internal strategic decisions. These issues can thus be regarded as key issues on an international level representing a global state of practice in sustainability management. On the other, in addition to these key issues, some companies manage further aspects relatively closely. These issues, however, are not or not yet established on a broad scale, since a large share of companies does not manage them very closely or does not even manage them at all.

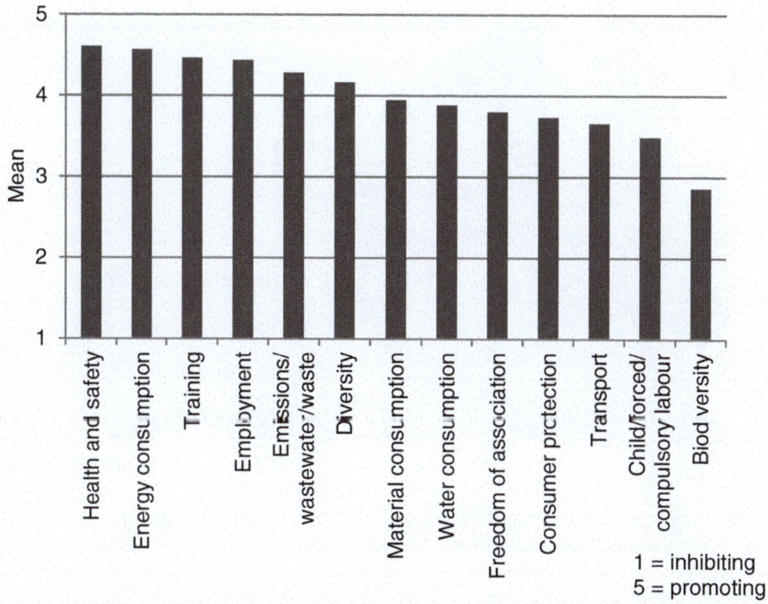

Fig. 3.2 Managed sustainability issues (n ranging from 442 to 463)
Question: "Which of the following issues does your company manage?"

3.2.2 *Integration*

3.2.2.1 Linking Sustainability to Core Business

If the sustainability activities of a company are to go beyond philanthropy and if the value creation processes are to become more sustainable, then sustainability must be linked to its core business activities so that economic success is created not alongside but *through* creating business cases for sustainability (Schaltegger and Burritt 2005; Schaltegger et al. 2012a). To investigate whether corporate sustainability practice is more or less integrated into their core business activities, the companies were asked to assess to what extent they link corporate sustainability to their core businesses.

More than two-thirds of all companies (see Fig. 3.3) state that they link sustainability at least to most segments of their core businesses, whereas only a minority of 7.3 % responds that they do not integrate sustainability at all or only to a few segments of their core businesses.

To substantiate their rating, the companies were asked to give examples for this integration. Whereas some examples (e.g. philanthropy) do not provide plausible proof that sustainability is linked with core economic activities, most examples (e.g. integrating sustainable products or services into the range) do illustrate that there is a degree of integration.

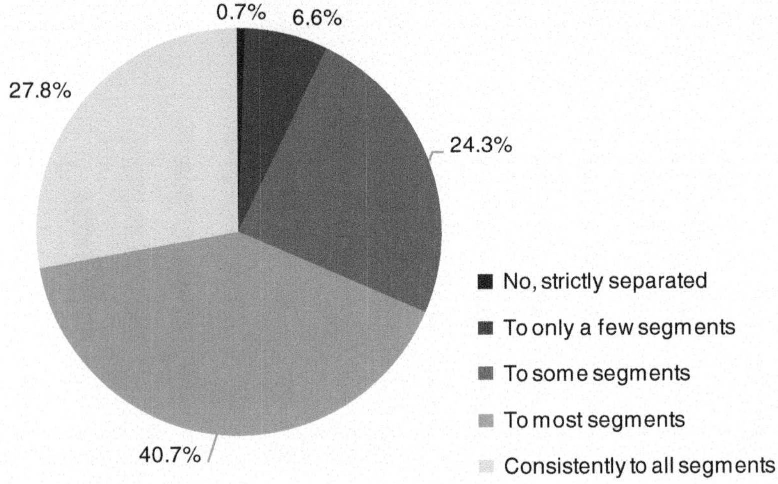

Fig. 3.3 Share of companies integrating sustainability into the core business (n = 457) Question: "Is the sustainability commitment of your company connected to its core business?"

3.2.2.2 Involvement of Organisational Units

Similar to the integration of sustainability with the core business, the involvement of all organisational units is considered to be essential to the successful management of many sustainability tasks (Gattiker and Carter 2010; Schaltegger et al. 2014). Likewise, Shrivastava and Hart (1995) consider cross-functional collaboration as a necessary condition for integrating sustainability into core economic activities.

Whereas it is commonly assumed that organisational units such as PR are involved (Clark 2000; Black and Härtel 2004), other organisational units such as finance are not frequently associated with sustainability measures. It is therefore questionable whether corporate practice is able to involve all functional units in sustainability measures to a significant degree (see Shrivastava and Hart 1995; Gattiker and Carter 2010).

To address this question, the company representatives were asked to assess each functional unit's engagement with sustainability measures on a five-point semantic differential scale (see Fig. 3.4). The results reveal large differences in sustainability engagement between organisational units. Unsurprisingly, CSR/sustainability departments are by far the most comprehensively engaged, followed by top management and public relations/corporate communication. At the other end of the scale, finance as well as financial and management accounting departments are least engaged in sustainability management. Interestingly, the differences between companies are relatively high for all organisational units with the exception of CSR/sustainability (as indicated by standard deviations of at least 1.00). The results therefore reveal not only large differences in the engagement of different functional units but also in the extent to which different companies involve their functional units.

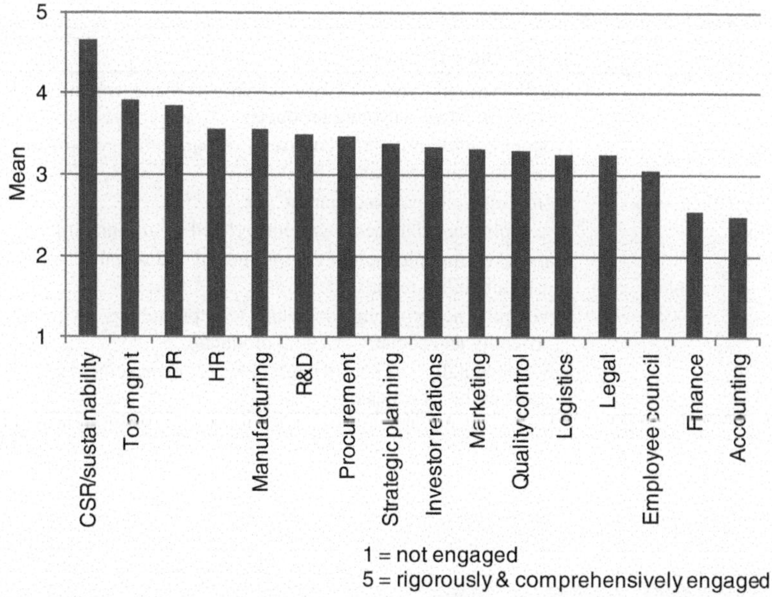

Fig. 3.4 Engagement with sustainability of organisational units (n ranging from 286 to 418) Question: "Which of the following organisational units/internal stakeholders engage with sustainability measures in your company?"

3.2.2.3 Business Case Drivers for Sustainability

Integrative sustainability management requires the creation of business cases for sustainability. In other words, instead of creating and redistributing profits created in an unsustainable manner, the company's whole value and profit creation process as such has to be sustainable. This, in turn, is expressed in corporate environmental and social measures which contribute to the company's economic success, since sustainability measures which are managed separately from a company's economic activities are most likely to be eliminated in times of crises (Schaltegger et al. 2012a). Whereas connecting sustainability to core business activities and involving all organisational units can be regarded as general conditions for realising sustainability profits, specific business cases for sustainability can be addressed through different drivers. The literature identifies seven distinct drivers for business cases of corporate sustainability (Olve et al. 1999; Schaltegger 2011). These are costs and efficiency, employee satisfaction, innovation, reputation, revenue and risk control. To analyse the relevance of these drivers in practice, each driver was matched with a corresponding measure of sustainability management (see Table 3.3).

The companies surveyed could indicate on a five-point semantic differential scale to what extent each measure is applied in their company (see Fig. 3.5). Efficiency, risk control and reputation are the drivers most frequently addressed, whereas

Table 3.3 Drivers of business cases for sustainability

Business case drivers	Sustainability measures
Costs	Environmentally and socially-oriented cost management (e.g. using cost-effective recycling products)
Efficiency	Producing with more efficient use of resources (e.g. optimising production processes)
Employee satisfaction	Promoting employee motivation
Innovation	Developing new business segments related to sustainability
Reputation	External communication of environmental and social activities (e.g. sustainability reporting)
Revenue	Developing new customer segments (e.g. promoting environmentally friendly and socially oriented products)
Risks	Environmentally and socially-oriented risk management (e.g. health care at the workplace)

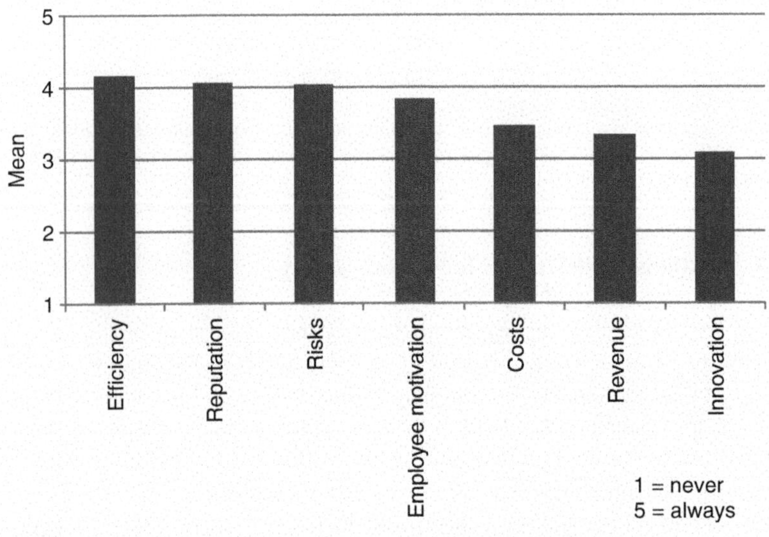

Fig. 3.5 Drivers of business cases for sustainability (n ranging from 397 to 405)
Question: "A number of sustainability measures are listed below. Are these implemented (in part) in your company?"

innovation and revenue are least frequently addressed. Again, relatively high standard deviations (1.22 and 1.18) indicate that these less commonly addressed drivers are of high importance in some companies, whereas other companies never address them. The standard deviation for the most important drivers, risk and efficiency, is considerably lower, which suggests that these drivers are established in corporate sustainability management practice.

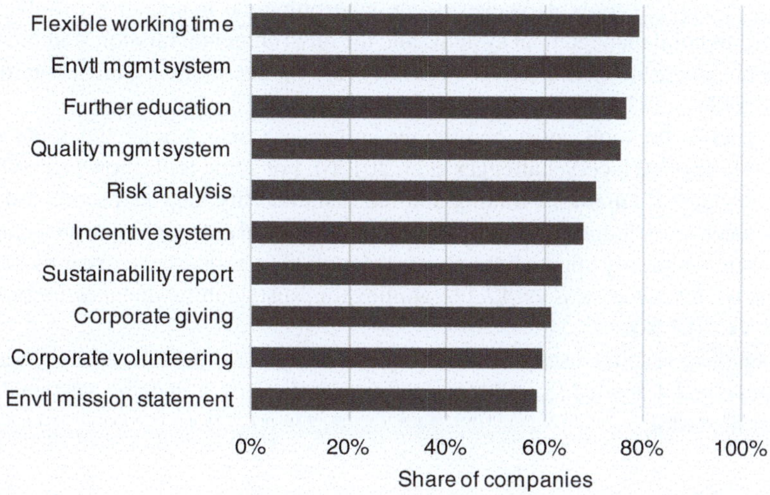

Fig. 3.6 The ten most commonly applied sustainability management tools (n ranging from 467 to 468)
Question: "Which methods of sustainability management [. . .] are applied in your company?"

3.2.3 Implementation

3.2.3.1 Sustainability Management Tools

Numerous management tools have been developed and promoted to help companies to implement corporate sustainability (e.g. Schaltegger et al. 2002, 2012b; Hahn and Scheermesser 2006). These tools address either environmental (e.g. environmental management systems, green purchasing) or social issues (e.g. social mission statement, social/fair labels) or deal with sustainability issues in an integrative manner (e.g. sustainability report, sustainable supply chain management). In addition to these tools specifically developed for corporate sustainability management, some conventional management tools such as quality management systems or incentive systems can be used in the context of sustainability.

To examine which tools are commonly applied in corporate practice, a list of 79 tools was drawn up based on academic and practitioner-oriented publications. For each of these tools, the participants could indicate whether they know of it and whether they apply it. Additionally, participants had the opportunity to add further tools which were not included in the list provided.

As Fig. 3.6 displays, many conventional management systems can be found among the most commonly used tools (e.g. flexible working time, further education and quality management system). While many of these conventional tools are related to HR/personnel issues (e.g. flexible working time, further education), no broader society-oriented management tool can be found among the ten most commonly applied tools. However, some environmental management tools, such as environmental management systems and the environmental mission statement, as well as one integrative tool, the sustainability report, rank in the ten most commonly applied tools.

In general the tools most commonly known are also among the tools most commonly applied (see Schaltegger et al. 2012b). For some tools, however, such as company-internal emissions trading and the sustainability balanced scorecard, this link between knowledge and application cannot be found in the international dataset, as they are commonly known but only rarely applied (e.g. internal emissions trading scheme: 61.7 % awareness vs. 17.1 % application; sustainability balanced scorecard 65.1 % vs. 20.3 %).

Given these results, research is challenged to examine what other factors, besides awareness and knowledge, influence the application of a specific sustainability management tool.

3.2.3.2 Sustainability Standards and Guidelines

Due to the growing importance of sustainability management, sustainability-related standards and guidelines, such as ISO 14000, EMAS or GRI, have been introduced (Elefsiniotis and Wareham 2005; Bracke and Albrecht 2007). By surveying the awareness and application of a list of 12 standards and guidelines, the ICSB investigates which of these standards and guidelines are commonly adopted in corporate practice (see Fig. 3.7).

The results demonstrate that, with regard to both awareness and application, ISO 9000 and ISO 14001 are the most popular standards. The awareness of ISO 26000 is at a similarly high level, but it is so far applied in no more than 15.4 % of the companies. The GRI guidelines have an awareness level of 80 % and are used by 53 % of the companies. All other standards and guidelines are applied by less than 35 % of the companies, with SA 8000, EFQM and AA 1000 being least applied.

3.3 Conclusion

The results of the ICSB survey suggest that corporate sustainability management is not only of academic interest but that companies in different regions of the world are working on implementing sustainability measures and on integrating them into their core economic activities. This becomes apparent in all three perspectives analysed in the ICSB, namely the intention, the integration and the implementation of corporate sustainability management.

In their study on CEO reflections on corporate sustainability, Lacy et al. (2010) argue that consumers are or will become the most important drivers of sustainability efforts. While the ICSB also reveals a positive influence of consumers, it identifies other stakeholders such as NGOs and media/public as even more promoting.

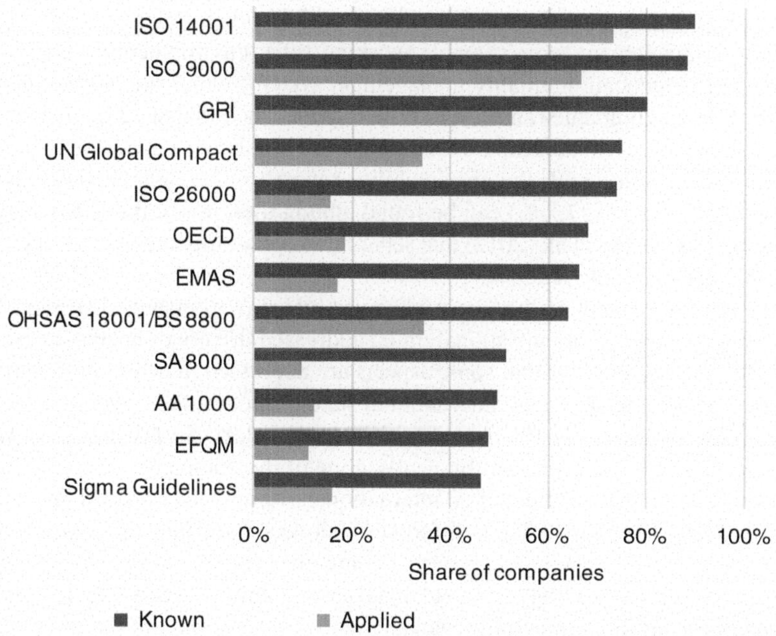

Fig. 3.7 Awareness and application of sustainability management standards and guidelines ($n_{INT} = 468$)
Question: "Which standards/guidelines are known in your company and which are applied?"

Similar to the ICSB, which identifies banks and creditors as the least promoting stakeholders, Lacy et al. (2010: 11) find "the investment community" to be a major obstacle to corporate sustainability engagement. Similar patterns can be observed concerning the importance of different sustainability issues for corporate practice. Lacy et al. (2010) identify education and climate change as the two most critical sustainability issues. The ICSB confirms these findings, with further education and energy consumption being among the three most important sustainability issues companies manage.

For the integration of sustainability issues, the ICSB finds that about two-thirds of all companies investigated connect sustainability to at least most segments of their core businesses. This finding thus confirms the study by Lacy et al. (2010: 33), where an increasing share of 81 % of CEOs stated that "sustainability issues are fully embedded into the strategy and operations" of their companies. One reason for this development may be the high importance of sustainability for top management, which was identified by Kiron et al. (2013) and is confirmed by the ICSB data.

Lacy et al. (2010) identify reputation as the main motivator for corporate sustainability and emphasise the generally defensive orientation of most corporate sustainability strategies. The ICSB, which investigates the extent to which each driver of business cases for sustainability is used, provides support for these findings. More specifically, it finds that the more defensive drivers of reputation, risk control and efficiency are most frequently addressed with sustainability activities.

With regard to actual implementation, the ICSB identifies a number of sustainability management tools (flexible working time, environmental management system, further education, quality management system) which are applied in more than 75 % of all companies and can thus be regarded as established in international corporate practice. In addition, many tools are applied in more than half of the companies. Since both the sustainability report (63.2 % application) and the environmental report (57.3 %) can be found among these tools, the ICSB supports the finding of Haanaes et al. (2012) that separate sustainability reporting is relatively widespread in corporate practice.

In general, for many of the questions investigated (e.g. managed sustainability issues, involvement of organisational units, addressed drivers of business cases for sustainability), we observe that some aspects are addressed in nearly all companies and can thus be considered as 'international state of the art' (e.g. managing energy consumption or involving PR in sustainability activities). Surprisingly, we rarely find aspects which are addressed by nearly none of the companies. Instead, for the aspects which are least addressed on international average, the standard deviation is relatively high. This signals large differences between the companies surveyed and thus suggests that a considerable group of companies chooses to manage more than just the internationally widely acknowledged sustainability aspects which constitute the state of the art in sustainability management.

The following chapters on the national results will reveal whether companies from some countries exceed this state of practice more frequently than others or whether some countries have not yet adopted some aspects of the international standard.

References

Berns M, Townsend A, Khayat Z, Balagopal B, Reeves M, Hopkins MS, Kruschwitz N (2009) The business of sustainability. Findings and insights from the first annual business of sustainability survey and the Global Thought Leaders' research project. MIT Sloan Management Review, Massachusetts Institute of Technology/Boston Consulting Group, North Hollywood

Black LD, Härtel CEJ (2004) The five capabilities of socially responsible companies. J Public Aff 4(2):125–144

Bracke R, Albrecht J (2007) Competing environmental management standards: how ISO 14001 outnumbered EMAS in Germany, the UK, France, and Sweden. Environ Plan C Govern Pol 25(4):611–627

Clark CE (2000) Differences between public relations and corporate social responsibility: an analysis. Public Relat Rev 26(3):363–380

Darnall N, Henriques I, Sadorsky P (2010) Adopting proactive environmental strategy: the influence of stakeholders and firm size. J Manag Stud 47(6):1072–1094

DiMaggio PJ, Powell WW (1983) The iron cage revisited: institutional isomorphism and collective rationality in organizational fields. Am Sociol Rev 48(2):147–160

Ditlev-Simonsen CD, Midttun A (2011) What motivates managers to pursue corporate responsibility? A survey among key stakeholders. Corp Soc Responsib Environ Manag 18(1):25–38

Elefsiniotis P, Wareham DG (2005) ISO 14000 environmental management standards: their relation to sustainability. J Prof Issues Eng Educ Pract 131(3):208–212

Freeman RE (1984) Strategic management: a stakeholder approach. Cambridge University Press, Cambridge
Gattiker TF, Carter CR (2010) Understanding project champions' ability to gain intra-organizational commitment for environmental projects. J Oper Manag 28(1):72–85
Haanaes K, Arthur D, Balagopal B, Kong MT, Reeves M, Velken I, Hopkins MS, Kruschwitz N (2011) Sustainability: the 'embracers' seize advantage. MIT Sloan Management Review, Massachusetts Institute of Technology/Boston Consulting Group, North Hollywood
Haanaes K, Reeves M, von Streng Velken I, Audretsch M, Kiron D, Kruschwitz N (2012) Sustainability nears a tipping point. MIT Sloan Management Review, Massachusetts Institute of Technology/Boston Consulting Group, North Hollywood
Hahn T, Scheermesser M (2006) Approaches to corporate sustainability among German companies. Corp Soc Responsib Environ Manag 13(3):150–165
Idowu SO, Leal Filho W (eds) (2009) Global practices of corporate social responsibility. Springer, New York/London
Kiron D, Kruschwitz N, Haanaes K, Reeves M, Goh E (2013) The innovation bottom line. How companies that see sustainability as both a necessity and an opportunity, and change their business models in response, are finding success. MIT Sloan Management Review/Boston Consulting Group, Boston
Lacy P, Cooper T, Hayward R, Neuberger L (2010) A new era of sustainability. CEO reflections on progress to date, challenges ahead and the impact of the journey toward a sustainable economy. http://www.accenture.com/SiteCollectionDocuments/PDF/Accenture_A_New_Era_of_Sustainability_CEO_Study.pdf. Accessed 8 May 2014
Lee M (2011) Configuration of external influences: the combined effects of institutions and stakeholders on corporate social responsibility strategies. J Bus Ethics 102(2):281–298
Olve N, Roy J, Wetter M (1999) Performance drivers. Wiley, Chichester
Sarkis J, González-Torre P, Adenso-Díaz B (2010) Stakeholder pressure and the adoption of environmental practices: the mediating effect of training. J Oper Manag 28(2):163–176
Schaltegger S (2011) Sustainability as a driver for corporate economic success. Consequences for the development of sustainability management control. Soc Econ 33(1):15–28
Schaltegger S, Burritt R (2005) Corporate sustainability. In: Folmer H, Tietenberg T (eds) The international yearbook of environmental and resource economics 2005/2006. Edward Elgar, Cheltenham, pp 185–222
Schaltegger S, Herzig C, Kleiber O, Müller J (2002) Sustainability management in business enterprises. Concepts and instruments for sustainable organisation development. The Federal Ministry for the Environment, Nature Conservation and Nuclear Safety (BMU)/Federation of German Industries (BDI)/Centre for Sustainability Management (CSM), Bonn/Berlin/Lüneburg
Schaltegger S, Lüdeke-Freund F, Hansen EG (2012a) Business cases for sustainability: the role of business model innovation for corporate sustainability. Int J Innov Sustain Dev 6(2):95–119
Schaltegger S, Windolph SE, Herzig C (2012b) Applying the known. A longitudinal analysis of the knowledge and application of sustainability management tools in large German companies. Soc Econ 34(4):549–579
Schaltegger S, Harms D, Hörisch J, Windolph SE, Burritt R, Carter A, Truran S, Crutzen N, Ben Rhouma A, Csutora M, Tabi A, Kokubu K, Kitada H, Haider MB, Kim J-D, Lee K-H, Moneva JM, Ortas E, Álvarez-Etxeberria I, Daub C-H, Schmidt J, Herzig C, Morelli J (2013) International corporate sustainability barometer: a comparative study of 11 countries. Centre for Sustainability Management, Lüneburg
Schaltegger S, Harms D, Windolph SE, Hörisch J (2014) Involving corporate functions: who contributes to sustainable development? Sustainability 6(5):3064–3085
Sharma S, Henriques I (2005) Stakeholder influences on sustainability practices in the Canadian forest products industry. Strateg Manag J 26(2):159–180
Shrivastava P, Hart S (1995) Creating sustainable corporations. Bus Strateg Environ 4(3):154–165
World Bank (2012) World DataBank. http://databank.worldbank.org/data/views/reports/tableview.aspx?isshared=true&ispopular=series&pid=2. Accessed 17 Sept 2013

Part II
Country-Specific Findings

Chapter 4
International Corporate Sustainability Barometer – Australia

Roger Burritt and Amanda Carter

Abstract Agribusiness and mining activities for which Australia is well known are not the industries of choice for contemplating a sustainability barometer. Agribusiness uses and exports much water in its products, from the driest inhabited continent on earth. Mining removes and is heavily challenged to replace biodiversity, ecosystems, and geological structures. However, the need for sustainability management using available tools is strengthened by this very focus. Value added from the Corporate Sustainability Barometer is at a premium in these circumstances. Results reveal these differences for Australia, including the stronger focus on water management. It is sad to report that Australia lags international practice in sustainability management and companies are less aware of available tools. The Barometer and its results for corporate Australia provide a foundation for stakeholders, both internal and external, to focus on promotion, training, awareness development and incentive based policy setting towards future improvement of sustainability management.

R. Burritt (✉)
Faculty of Business and Economics, Department of Accounting and Corporate Governance, Macquarie University, Sydney, Australia
e-mail: roger.burritt@mq.edu.au

A. Carter
Centre for Accounting, Governance and Sustainability, School of Commerce, University of South Australia, Adelaide, Australia
e-mail: amanda.carter@unisa.edu.au

S. Schaltegger et al. (eds.), *Corporate Sustainability in International Comparison*, Eco-Efficiency in Industry and Science 31, DOI 10.1007/978-3-319-06227-3_4,

© Springer International Publishing Switzerland 2014

4.1 Introduction

4.1.1 The Australian Context

Australia is a country that is well known for its hot, sunny weather, preponderance of desert landscapes and Aboriginal and Torres Strait Islander cultural attributes. However, in terms of corporate activity the country is similarly well known for its strong primary sector, with particular importance in the mining (Battellino 2010) and agricultural (Connolly and Lewis 2010) industries. It has a small secondary sector, including a small manufacturing industry, but has vibrant banking, tourism and higher education industries in the tertiary sector (Abbott et al. 2013). Sustainability is widely understood in the context of triple bottom line reporting in Australia (Frost et al. 2005; Golob and Bartlett 2007; Herbohn et al. 2012). Sustainability issues that have risen to particular prominence include carbon emission issues as reporting is required by the *National Greenhouse and Energy Reporting* (*NGER*) *Act*, water shortages and associated water efficiency as Australia is the driest continent in the world, and rehabilitation by mining companies, an ongoing issue in Australia for many decades.

4.1.2 The Australian Sample

Australian companies included in the International Corporate Sustainability Barometer survey coordinated by the Centre for Sustainability Management (see Schaltegger et al. 2013) were selected from the Business Review Weekly (BRW) Top 1000 list (2011). Such companies were also listed on the Australian Stock Exchange. Contact names for each company were obtained and a phone call made to establish the initial link. Each company was approached in early 2012 to complete the survey that was available through a website. When each company was contacted by phone and email it was offered assistance to complete the survey. Only three companies requested such assistance. The final 183 companies were provided with the link to the survey. A total of 55 completed surveys were received. The final Australian sample for the data collection was reduced to 48 companies as a requirement for international comparison was that each company should be earning more than €50 million. In summary, 48 companies were included in the final data set of Australian companies and the response rate was 26.2 %.

The size profile of Australian companies responding indicated that 79.2 % employed up to 10,000 employees with an additional 18.8 % employing between 10,001 and 100,000 people. Only one respondent employed more than 100,000. With 53.8 % of the international group employing up to 10,000 people, the Australian companies generally employed a lower number of employees than their international counterparts (see Fig. 4.1).

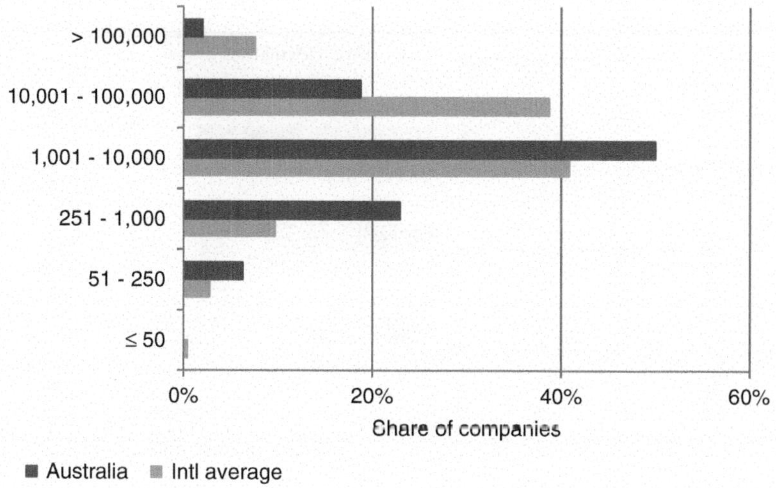

Fig. 4.1 Number of employees ($n_{AUS} = 48$, $n_{INT} = 465$)

A second measure of size of companies responding is sales turnover. Around 31 % of Australian companies reported revenues of between €50 million and €500 million compared with 13.9 % for the international group. No Australian company reported revenues in excess of €50,000 million whereas around 10 % of the international group did so. Hence, sales revenues tended to be lower than the international average. Non-domestic (i.e. overseas) sales as a proportion of total revenues were much smaller for Australian companies than for the international group: with 33.3 % reporting only up to 1 and 31.1 % reporting between more than 1 and 20 % of total revenues falling in the non-domestic category. The international group was more evenly distributed with around 13 % having 60–80 % of total revenues as non-domestic sales and 15.4 % with more than 80 % of revenues from non-domestic sales. Hence, in relation to employees, sales revenues and non-domestic sales the Australian companies were lower than the international sample of companies examined.

The core businesses of the Australian group were identified as being stronger in commodities, auxiliary materials, energy, chemical and pharmaceutical industries than the international group but lower in consumer goods, trade and logistics (see Fig. 4.2). Hence, in addition to the importance of the primary sector the secondary sector, through chemicals and pharmaceutical industries, is relatively more important than globally. Given the large area of the country and a strong export focus, it is not a surprise to also find logistics and trade are relatively important industries.

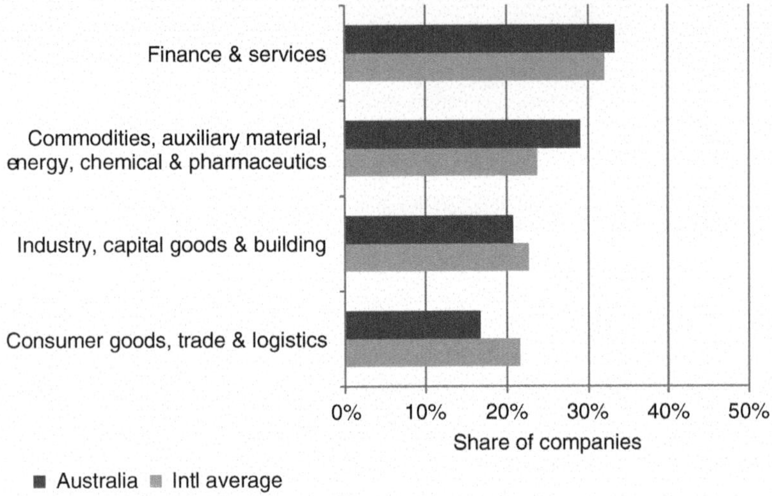

Fig. 4.2 Core business areas ($n_{AUS} = 48$, $n_{INT} = 468$)

4.2 Analysis: Comparison of Australian Companies with the International Sample

The survey asked questions regarding the role internal and external stakeholders had in terms of the impact on decisions to adopt a sustainability approach to business and the issues that were managed (Schaltegger et al. 2003). It sought to discover the intention of stakeholders towards adopting sustainability tools, how well sustainability was integrated into core business and how systematically available sustainability tools were applied. An overview of the results can also be found in the International Corporate Sustainability Barometer report (Schaltegger et al. 2013).

4.2.1 Intention

Different stakeholder groups require different policies from company management in order to address their concerns and interests. Hence, one key matter for companies is whether stakeholders are seen to be promoting sustainability or inhibiting sustainability as a core aspect of company business.

4.2.1.1 Motivation

The influence of external stakeholders on the sustainability approach used by companies was assessed in terms of whether they inhibited or promoted sustainability. Figure 4.3 shows for the international group all of the stakeholder groups identified

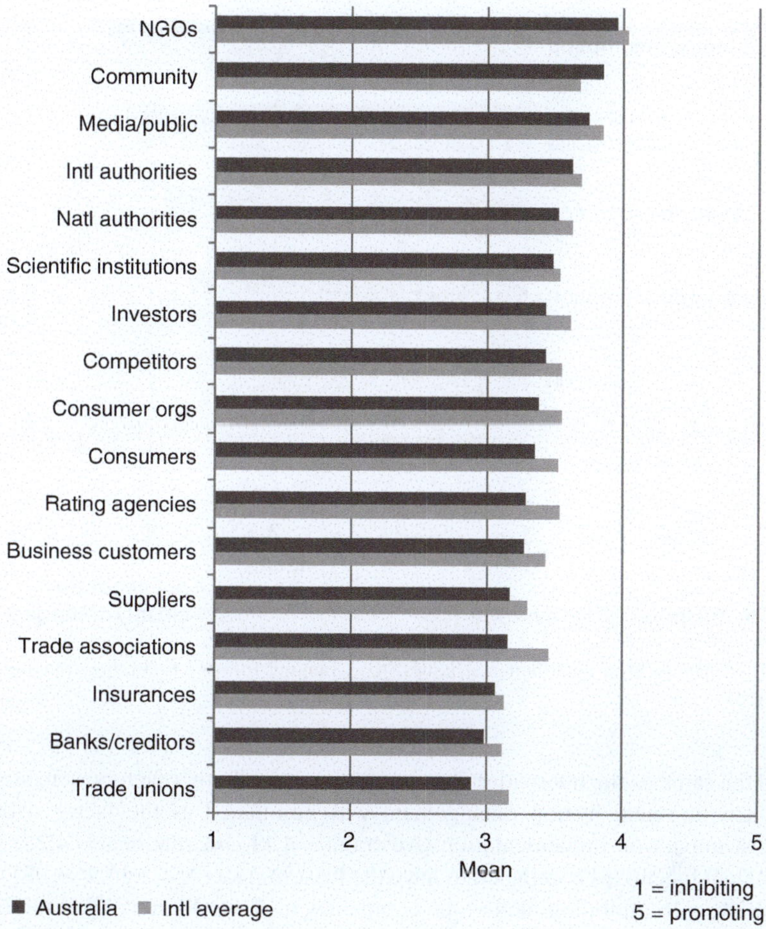

Fig. 4.3 Impact of external stakeholders (n_{AUS} ranging from 42 to 47, n_{INT} ranging from 393 to 450)

as having a promoting influence. Australian companies identified similar results. NGOs/environmental/social organisations were identified by both Australian companies and the international sample as having the strongest promoting influence, with community interest being second most important in terms of impact, something unique to the country in comparison with the global sample. Another difference was that Australian companies identified trade unions and banks/creditors as having a slightly inhibiting effect on the adoption of a sustainability approach, perhaps surprising for banks given their strong involvement in the United Nations Environment Program Finance Initiative (UNEP Financial Initiative 2013) for many years.

Other factors influencing motivation to implement a sustainability approach include a slightly higher relative impact on Australian companies from government incentives. Self-commitment by business, practicability of methods and legislation

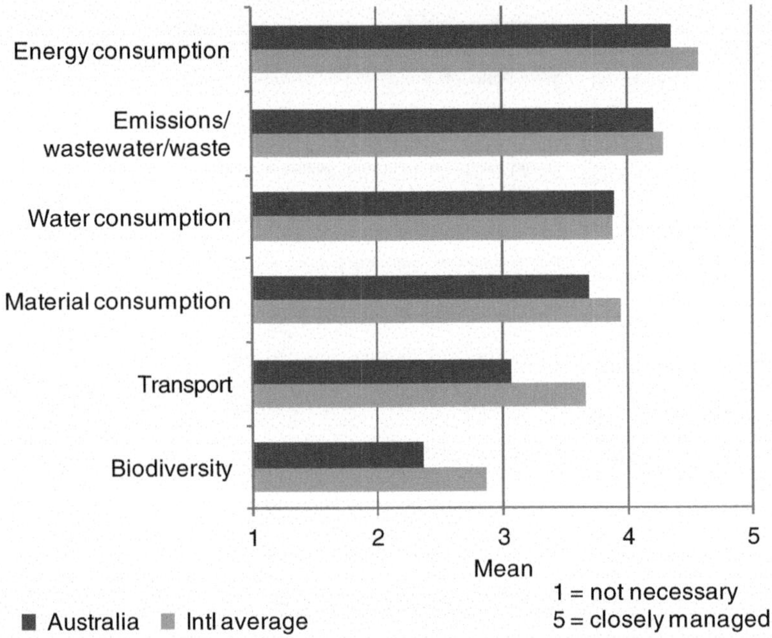

Fig. 4.4 Management of environmental issues (n_{AUS} ranging from 47 to 48, n_{INT} ranging from 455 to 463)

ranked the same as the international average, whereas all other factors were slightly lower. Commitment by top management was nominated as the factor with the strongest impact on implementation (Australia = 4.11, international = 4.28), with government incentives nominated to have the least impact (A = 3.45, i = 3.31).

In relation to inhibiting factors the companies in Australia ranked all inhibitors as having higher impacts than the international sample. The international group of respondents identified lack of personnel capacities as the inhibiting factor with the strongest impact (3.11) on adoption of a sustainability approach, whereas the Australian respondents indicated a lack of government incentives (3.58) to be the strongest. The inhibiting factors with the least impact were self-commitment by business (international = 2.56) and legislation (Australia = 2.98).

4.2.1.2 Issues

From the range of environmental issues included in this survey, energy consumption was the most closely managed by both the Australian and the international group (see Fig. 4.4).

Emissions, wastewater and waste were a close second. This is perhaps unsurprising because of the link between energy and emissions. Biodiversity ranked as the issue least managed by both groups, with Australian companies indicating a much

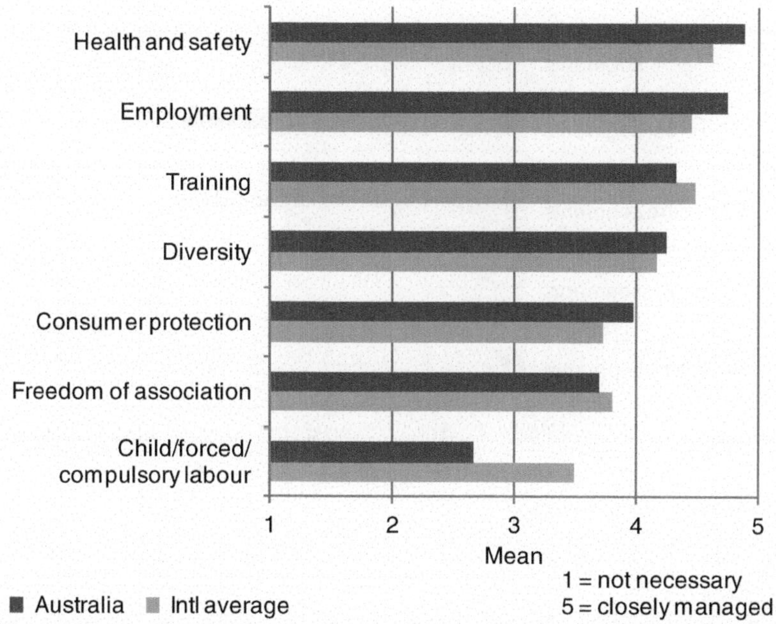

Fig. 4.5 Management of social issues (n_{AUS} ranging from 45 to 47, n_{INT} ranging from 442 to 461)

lower level of management than the international group. This is in contrast to the strength of Australia's biodiversity and its management at national and subnational levels including the incorporation of traditional environmental management techniques (see for example Butler et al. 2012; Lindenmayer et al. 2012; Taylor et al. 2013 and Natural Resource Management Ministerial Council 2010 for a 20-year strategy). Terminology may play a part in this perception as management of natural resources is sometimes used as a proxy for biodiversity by players in the Australian setting.

Distinct differences exist between the Australian and international respondents with regard to the social issues being managed. Australian companies more closely managed issues of occupational health and safety, workplace and employment, diversity and equal opportunity, as well as consumer protection than the international respondents (see Fig. 4.5). As a single issue, occupational health and safety was the issue with the closest management by companies in both groups (Australian mean 4.87 and international 4.61), values for environmental issues were all lower too, whereas child, forced and compulsory labour (e.g. in the supply chain) was scored lowest in both groups (Australian mean = 2.67 and international = 3.49).

Perceptions of issues that stakeholders required to be managed by companies vary between the two groups. Water-related issues score higher in Australian responses with energy consumption receiving the highest score for international stakeholder expectations (3.77). Australian companies specifically identified

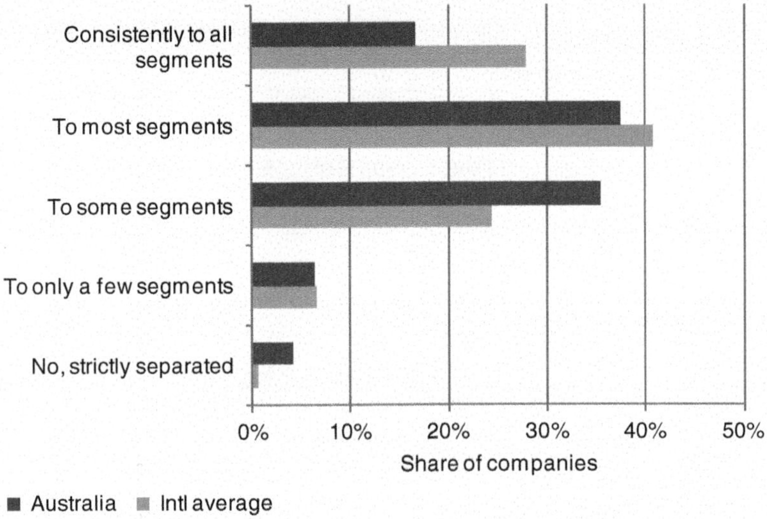

Fig. 4.6 Connection of sustainability to core business ($n_{AUS} = 48$, $n_{INT} = 457$)

emissions/wastewater/waste as the most important issue (3.77). In both groups biodiversity is the environmental issue least expected to be managed (A = 2.20, i = 2.55).

4.2.2 *Integration*

4.2.2.1 Connection to Core Business

For a small group of Australian respondents (4.2 %), sustainability and core business were strictly separated within their companies, compared with just 0.7 % of the international group (see Fig. 4.6). However, most Australian and international respondents made some attempt at connecting core business to sustainability, whether some, most or all segments. Compared with the international group, the Australian companies tended to have less segments connected to sustainability, although around 17 % did have all segments of core business connected.

4.2.2.2 Involvement of Organisational Units

Marketing and the employee council were ranked as units with higher impacts on sustainability effort relative to the international average but only marginally so (i.e. less than 0.07 difference between the means). CSR/sustainability (including environment, health, occupational safety) was ranked as the organisational unit with

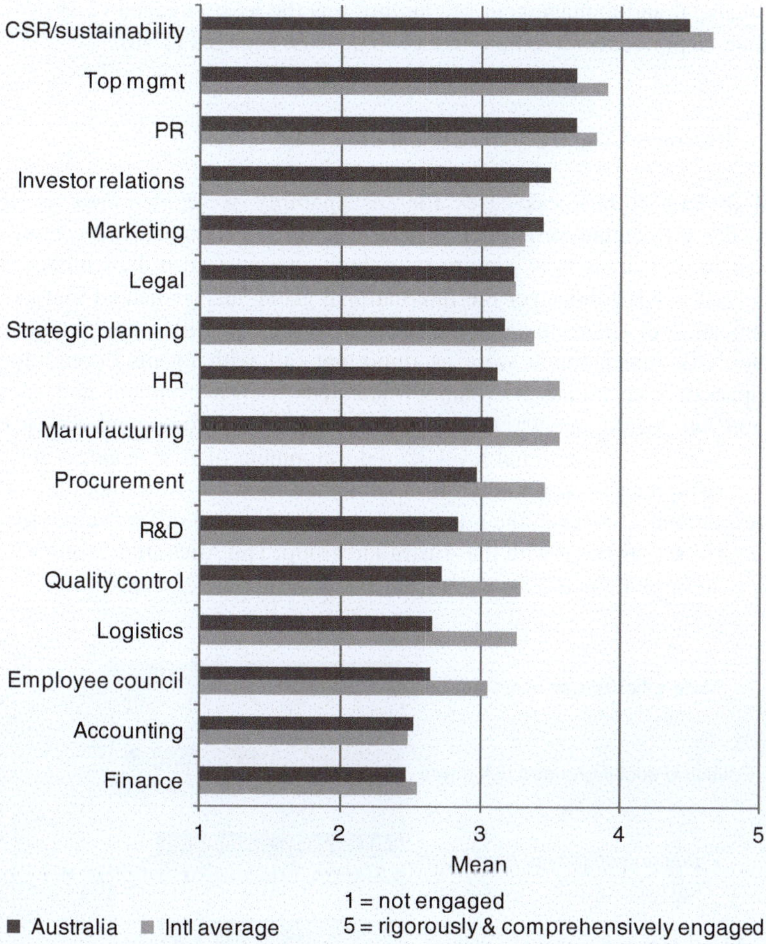

Fig. 4.7 Engagement with sustainability measures (n_{AUS} ranging from 17 to 46, n_{INT} ranging from 286 to 4)

the highest impact both internationally (4.76) and in Australia (4.64). The second highest organisational unit impact differed between the groups with international nominating top management and Australian companies indicating public relations. The organisational unit with the least impact was financial and management accounting (3.19) for the international sample and logistics and distribution (3.00) for the Australian sample of companies.

Perhaps unsurprisingly, corporate social responsibility or sustainability organisational units were more rigorously and comprehensively engaged than any other (Australian mean 4.49, international 4.66; see Fig. 4.7). Top management was the second most engaged for both. The Australian companies identified public relations/corporate communication as equal second. For the international group financial and management accounting was the least engaged (2.49) whereas Australian respondents identified finance to be least engaged (2.48).

4.2.2.3 Business Case Drivers for Sustainability

Several drivers of business cases for sustainability were identified as important by the Australian respondents (see Fig. 4.8). Producing with more efficient use of resources was dominant in both groups (Australian mean = 4.09, international = 4.16), however the international group also indicated that external communication of environmental and social activities and environment and social-orientated risk management were as important (all with means exceeding 4.0). In comparison, Australian companies indicated that environment and socially-orientated risk management was also important but was implemented to a lesser degree than at the international level. The development of new business segments related to sustainability was the least implemented by both Australian and international respondents, however, the international group (mean 3.09) was more likely to implement this, compared with the Australian group. The Australian group was also less likely to implement the development of new customer segments.

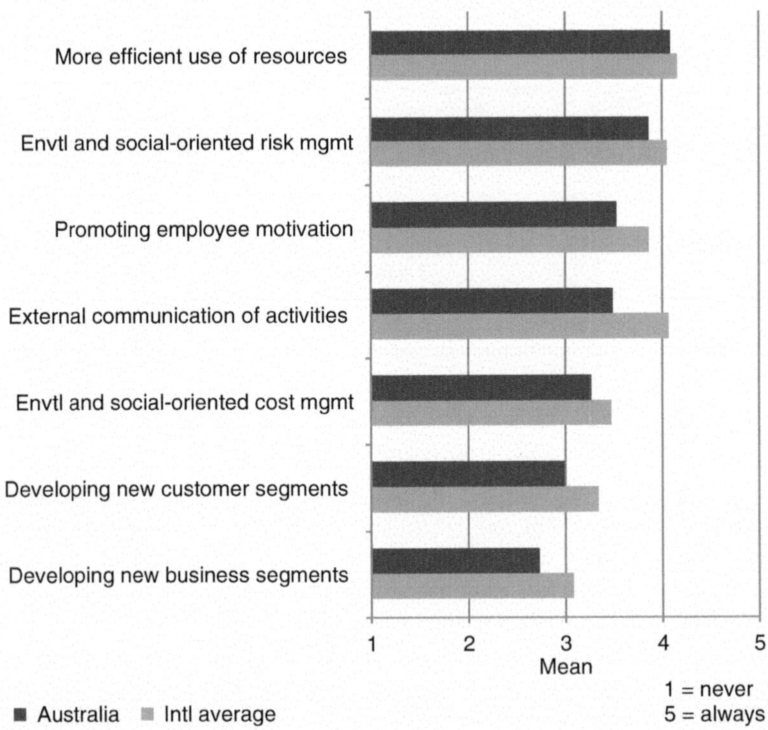

Fig. 4.8 Implementation of sustainability measures (n_{AUS} ranging from 46 to 48, n_{INT} ranging from 397 to 405)

4.2.3 *Implementation*

4.2.3.1 Stakeholder Management

Stakeholder management through observing stakeholders in general is not a popular choice but is used by more international than Australian companies sampled. In both groups it is used on a case-by-case basis by about 48 % with about 3 % more companies in the international sample using observation on a more general basis (A = 40.4 %, i = 43.5 %). Management through an information strategy (e.g. through reports, websites, the press, etc.) is used by 77.1 % of international respondents on a general basis (63.0 % of the Australian sample). Dialogue with stakeholders/seeking advice (e.g. questionnaires, dialogue forums) is used more by the international group, where over 50 % use it on a case-by-case basis (A = 44.7 %). However, 19.1 % of Australian respondents are not using this method (i = 10.1 %). Involvement in decision-making processes (e.g. in advisory boards) is used on a case-specific basis more by the international sample (53.4 %) than by Australian companies (46.8 %). Cooperating and networking to develop joint solutions (e.g. in working groups) is popular on a case-specific basis (61.1 % international, 53.2 % Australian), however, a large proportion of the Australian sample (31.9 %) and 16.1 % of the international sample do not use this method. Empowerment, such as through the provision of financial support for stakeholder management, is not used by around 40 % of each group, but with 49.5 % of the international group and 61.7 % of the Australian companies using it on a case-specific basis. Delegating decision-making authority to stakeholders is used by under 50 % in both groups. Only 32.9 % of the international sample and 38.3 % of the Australian company sample use this on a case-specific basis.

4.2.3.2 Sustainability Management Tools

A range of sustainability management tools were offered in the survey to gauge whether they were known and applied. Around half of the tools included in the survey were known *or* known and applied by 50 % or more of the companies in the Australian and international groups. Around 17 % of the tools listed were not known by international and Australian respondents (awareness below 50 %) including social cost accounting, sustainability accounting, and environmental shareholder value (see Table 4.1). A little less than half of the methods listed were unknown by Australian companies (48.1 %) including social management systems, eco-control, eco-audits, material and energy flow accounting, and eco-labelling. In the international group only one other tool was unknown, opportunity-risk dialogue.

With a range of guidelines available to companies, the survey asked which of these were known and/or applied. ISO 14001 was identified as the guideline most known in both groups (see Table 4.2), with 90 % of internationals and 88 % of Australian companies knowing this standard. Two additional ISO standards,

Table 4.1 Known sustainability management tools

	Australian responses less than 50 %	International responses less than 50 %	Both groups responses less than 50 %
ABC analysis (eco/social)			X
Checklists (eco/social)			X
Community advisory panel			
Corporate citizenship			
Corporate giving			
Corporate/employee volunteering			
Cross-impact-analysis			X
Early detection	X		
Eco-audit	X		
Eco-balance/life-cycle assessment			
Eco-benchmarking	X		
Eco-budgeting			X
Eco-circle	X		
Eco-compass			X
Eco-control	X		
Eco-design/design for environment			
Eco-efficiency analysis	X		
Eco-efficiency indicators	X		
Eco-indicators	X		
Eco-investment accounting			X
Eco-label	X		
Eco-marketing	X		
Eco-sponsoring	X		
Emission certificate trading (in-company)	X		
Environmental accounting			
Environmental cost accounting	X		
Environmental information system			
Environmental management system (EMS)			
Environmental mission statement			
Environmental report			
Environmental shareholder value			X
Environmental statement			
Flexible working time			
Further education			
Green purchasing			
Green supply chain management			
Human resource control			
Human resource report			
Incentive system			
Material and energy flow accounting	X		
Material flow analysis			
Material flow cost accounting	X		
Opportunity-risk dialogue		X	

(continued)

Table 4.1 (continued)

	Australian responses less than 50 %	International responses less than 50 %	Both groups responses less than 50 %
Product carbon footprint (PCF)			
Product line analysis	X		
Proposal system	X		
Quality circle	X		
Quality management system (QMS)			
Risk analysis			
Scenario analysis			
Social accounting			X
Social audit	X		
Social benchmarking			X
Social cost accounting			X
Social indicators	X		
Social management system (SMS)	X		
Social marketing			
Social mission statement			
Social report			
Social/cultural sponsoring			
Social/fair label	X		
Socio-eco-efficiency-analysis			X
Socio-efficiency indicators			X
Stakeholder dialogue/forum			
Stakeholder value indicators			X
Sustainability accounting			X
Sustainability audit			
Sustainability balanced scorecard			
Sustainability benchmarking			
Sustainability control	X		
Sustainability indicators			
Sustainability label	X		
Sustainability management system			
Sustainability marketing			
Sustainability mission statement			
Sustainability report			
Sustainability sponsoring			
Sustainable design			
Sustainable supply chain management			

9000 and 26000, were the second and third most known guidelines. In terms of application rather than knowledge, ISO 14001 was most applied (73.1 %) by international respondents (followed by ISO 9000 and the GRI guidelines) whereas in Australia the most applied was ISO 9000 (58.3 %) followed by ISO 14001 and the GRI guidelines. EFQM is the guideline that Australian respondents were least familiar with, with 70.8 % identifying it as unknown (compared with 52.1 % of

Table 4.2 Guidelines known and applied

	International, %			Australia, %		
	Known	Known and applied	Not known	Known	Known and applied	Not known
ISO 14001	90	73	10	88	50	13
ISO 9000	88	67	12	85	58	15
ISO 26000	74	15	26	67	15	33
EMAS	66	17	34	42	4	58
OECD Guidelines	68	18	32	58	6	42
GRI	80	53	20	69	27	31
UN Global Compact	75	34	25	63	10	38
EFQM	48	11	52	29	2	71
AA 1000	50	12	50	40	15	60
OHSAS 18001/BS 8800	64	35	36	48	15	52
SA 8000	52	10	49	38	6	63
Sigma Guidelines	46	16	54	48	13	52

internationals). For the internationals the Sigma Guidelines represented the least known (53.6 % identifying as not known compared with 52.1 % of Australians). The relative awareness of selected guidelines shows that Australian companies tended to have less general knowledge, for example OHSAS 18001/BS 8800, EMAS and SA 8000 were known by more than 50 % of internationals but less than 50 % of Australian respondents.

4.2.3.3 Measurement

While all respondents tend to have high levels of commitment to measuring their impact on environmental and social issues, some issues were conspicuous for their limited impact measurement. Child, forced and compulsory labour was the only social issue in which less than 50 % of the Australian respondents and internationals measured their impact, while biodiversity and transport for environmental issues had limited measurement. In line with the management of social and environmental issues by companies, energy consumption and emissions had high levels of measurement for both groups (over 90 %) and occupation health and safety as well as workplace/employment in the social arena.

4.3 Conclusion

In reviewing the Australian data and comparing it with the international data set it is clear that the size (in terms of employees), revenues and proportion of non-domestic sales in revenues are much smaller in the Australian sample than in the international sample. Industries in which the Australian respondents are involved also differ from the international group. Australian respondents generally replicated the influence of external stakeholder groups, although in all instances international respondents tended to score external stakeholders as more promoting than Australian companies. Water consumption was the only environmental issue that Australian respondents score as more closely managed and is consistent with moves towards water accounting (Turner et al. 2010); otherwise the Australian data was closely aligned with that of the international respondents. Occupational health and safety was the dominant social issue amongst the Australian respondents as well as workplace and employment issues. Engagement with sustainability measures is generally less comprehensive than the international group. Awareness of sustainability management tools in the Australian sample of companies is less well formed than for international respondents. Australian companies were dedicated to sustainability (Frost et al. 2005; Herbohn et al. 2012) within their business but had less developed knowledge, and commitment was not comprehensive within the group of firms.

In summary, overall there is considerable similarity between the Australian and international samples in relation to the responses to the adoption of a sustainability approach to business, environmental and social issues raised and managed, connection of sustainability to core businesses and less segments connected to sustainability, and stakeholder management strategies and tools (see also Schaltegger and Burritt 2005). However, the size of the companies in the samples differs markedly, with the Australian companies being smaller on average (in terms of revenue) and the engagement with sustainability measures on average being lower.

References

Abbott M, Wu S, Wang WC (2013) The productivity and performance of Australia's major banks since deregulation. J Econ Finance 37(1):122–135

Battellino R (2010) Mining booms and the Australian economy. Reserve Bank of Australia Bulletin, March Quarter:63–69

Business Review Weekly (2011) Top 1000 companies. 27 Jan to 2 Mar, pp 30–46

Butler JR, Tawake A, Skewes T, Tawake L, McGrath V (2012) Integrating traditional ecological knowledge and fisheries management in the Torres Strait, Australia: the catalytic role of turtles and dugong as cultural keystone species. Ecol Soc 17(4):1–19

Connolly E, Lewis C (2010) Structural change in the Australian economy. Reserve Bank of Australia Bulletin, September Quarter:1–9

Frost G, Jones S, Loftus J, van der Laan S (2005) A survey of sustainability reporting practices of Australian reporting entities. Aust Account Rev 15(35):89–96

Golob U, Bartlett J (2007) Communicating about corporate social responsibility: a comparative study of CSR reporting in Australia and Slovenia. Publ Relat Rev 33(1):1–9

Herbohn K, Dargusch P, Herbohn J (2012) Climate change policy in Australia: organisational responses and influences. Aust Account Rev 22(2):208–222

Lindenmayer D, Wood J, Montague-Drake R, Michael D, Crane M, Okada S, Gibbons P (2012) Is biodiversity management effective? Cross-sectional relationships between management, bird response and vegetation attributes in an Australian agri-environment scheme. Biol Conserv 152:62–73

Natural Resource Management Ministerial Council (2010) Australia's biodiversity conservation strategy 2010–2030. Australian Government, Department of Sustainability, Environment, Water, Population and Communities, Canberra

Schaltegger S, Burritt R (2005) Corporate sustainability. In: Folmer H, Tietenberg T (eds) The international yearbook of environmental and resource economics 2005/2006. Edward Elgar, Cheltenham, pp 185–222

Schaltegger S, Burritt R, Petersen H (2003) An introduction to corporate environmental management: striving for sustainability. Greenleaf, Sheffield

Schaltegger S, Harms D, Hörisch J, Windolph SE, Burritt R, Carter A, Truran S, Crutzen N, Ben Rhouma A, Csutora M, Tabi A, Kokubu K, Kitada H, Haider MB, Kim J-D, Lee K-H, Moneva JM, Ortas E, Álvarez-Etxeberria I, Daub C-H, Schmidt J, Herzig C, Morelli J (2013) International corporate sustainability barometer: a comparative study of 11 countries. Centre for Sustainability Management, Lüneburg

Taylor RS, Watson SJ, Bennett AF, Clarke MF (2013) Which fire management strategies benefit biodiversity? A landscape-perspective case study using birds in mallee ecosystems of south-eastern Australia. Biol Conserv 159:248–256

Turner GM, Baynes TM, McInnis BC (2010) A water accounting system for strategic water management. Water Resour Manag 24(3):513–545

UNEP Financial Initiative (2013) UNEP Financial Initiative. http://www.unepfi.org. Accessed 30 Oct 2013

Chapter 5
The Case of Belgium

Nathalie Crutzen

Abstract Sustainability management is a relatively recent concept in Belgium, compared to other European countries like Germany or the UK. Prior studies focusing on sustainability management in Belgian companies show that, while considerable improvements have been made over the last decade, sustainability management in Belgium varies greatly (Louche et al., Belgium. In: Idowu SO, Leal Filho W (eds) Global practices of corporate social responsibility. Springer, Berlin/Heidelberg, pp 125–147, 2009; Business & Society Belgium, Le baromètre CSR. Business and society Belgium magazine, Business & Society Belgium, Brussels, 2011).

This chapter compares the practices of Belgian companies with the average of the International Corporate Sustainability Barometer. The analysis confirms that sustainability management is a living and evolving concept in Belgium, but it also shows that Belgian practices are commonly below the international average and sometimes uneven. For example, the perceived influence of stakeholders as well as the awareness and application of sustainability management tools are below the international average in Belgium.

However, even if the integration into core business varies greatly, a particularly high percentage of Belgian companies connect sustainability to most segments of their core business. This observation shows that Belgian companies are particularly interested in finding opportunities to link sustainability with their core business.

This chapter presents and discusses the major differences between Belgium and the international average as well as different possible explanations.

N. Crutzen (✉)
HEC-Management School of the University of Liege, Liege, Belgium
e-mail: ncrutzen@ulg.ac.be

S. Schaltegger et al. (eds.), *Corporate Sustainability in International Comparison*, Eco-Efficiency in Industry and Science 31, DOI 10.1007/978-3-319-06227-3_5,
© Springer International Publishing Switzerland 2014

5.1 Introduction

Compared to other European countries, sustainability management is a relatively new phenomenon in Belgium. In fact, most discussions refer to the term 'corporate social responsibility' (CSR) instead of 'sustainability management'.

In 1997, sustainability was explicitly addressed by the Belgian legislation for the first time as a legal framework for sustainable development was created. In April 2006, this was followed by a 'Reference Framework for CSR', which was supplemented by the 'CSR Action Plan' in 2007 (Mazijn and Gouzee 2007; Louche et al. 2009). In addition to these government initiatives, several networks and platforms were established, leading to an increased number of actors and to a multiplication of initiatives since the mid-1990s.

Recent studies on the state of the art and development of sustainability management in Belgium reveal that these efforts may have resulted in significant improvements in many aspects of sustainability management. However, these studies also find that the degree of corporate engagement for sustainability varies greatly among Belgian firms (Louche et al. 2009; Business & Society Belgium 2011).

After an overview of the context to sustainability management in Belgium (see Sect. 5.1.1) and a presentation of the sample (see Sect. 5.1.2), this chapter compares the practices of Belgian companies with the international average. Section 5.2 underlines some key observations and confirms findings made by Louche et al. (2009) that large Belgian firms often perform below the international average. The chapter ends with a discussion and a conclusion about the situation in Belgium in the area of sustainability management, about future prospects and potential paths for improving sustainability management in Belgium.

5.1.1 The Belgian Context

Belgium is a small Central European country characterised by an above-average population density (Louche et al. 2009). Its federal institutional structure is based on three language communities (Flemish, French and German speaking) and three geographical regions (Flanders, Wallonia and Brussels). While the federal government is responsible for matters requiring uniform national policies (e.g. social security or control of air pollution from mobile sources), the regional governments are responsible for the majority of societal issues and are of crucial importance for most sustainability issues, since they also set the environmental objectives and are responsible for implementing related policies (O'Brien et al. 2001; Louche et al. 2009).

Belgium is a highly industrialised economy characterised by a high gross national product (GNP), high exports per capita and a highly productive and skilled workforce. Similar to the other Benelux countries, it has an extensive transportation infrastructure. The Belgian corporate landscape consists mainly of small and medium-sized enterprises (SMEs), with most large corporations in Belgium belonging to foreign groups. Only few Belgian companies (such as AB Inbev, the world's largest brewery group) employ several thousand employees globally.

Like the other Benelux countries, Belgium is a continental European welfare state, where the state takes an active role in shaping the economic rules (Keman 2008). Therefore, it is unsurprising that government regulations also play an important role for CSR and sustainability management in Belgium. The first legal framework for sustainable development was created in 1997. Corporate aspects of sustainability were first regulated by law in 2006 in the 'Reference Framework for CSR'. In 2007 this framework was followed by a CSR Action Plan (Mazijn and Gouzee 2007; Louche et al. 2009).

Besides government actions, over the last two decades, the number of actors involved in sustainability, such as corporate networks and platforms, has significantly increased. For example, Business & Society Belgium has become a very influential business network.

In addition to legislation and corporate practice, scientific research and academic education have also greatly developed over the last decades (Louche et al. 2009). The status and progress of corporate sustainability and CSR in Belgium has been investigated in four major studies (Business & Society Belgium 2005, 2011; FEB 2007; Louche et al. 2009), three of which are primarily practitioner-oriented reports (Business & Society Belgium 2005, 2011; FEB 2007). These studies show that, even if many aspects of corporate sustainability have improved considerably, great disparities in the actual sustainability management practices of large Belgian firms still exist (Louche et al. 2009; Business & Society Belgium 2011).

One of the first studies addressing CSR in Belgium was carried out among 250 companies by the Belgian Enterprises Federation (FEB) in 2007. Of these companies, 90 % acknowledge that they should also consider social and environmental objectives instead of focusing solely on profit-making. However, the vast majority does not address these aspects at a strategic level. One central weakness the FEB study (2007) revealed is the management of stakeholder relations. Most companies only engage with stakeholders on an irregular basis and largely neglect some stakeholders. Additionally, the study shows that, while the awareness of international standards and guidelines is relatively high among the investigated companies, the implementation of these standards and guideline is far less common.

Similarly, Louche et al. (2009) identify weaknesses concerning the actual implementation of CSR measures in Belgium. Based on secondary data provided by a French rating agency (Vigeo), they describe the Belgian state of the art in CSR within an international context. Concerning most of the issues investigated, Louche et al. (2009) find that Belgian companies do not perform as well as the international average. Additionally, they emphasise that the state of the art of CSR in Belgium is improving more slowly than in most other countries.

The most recent study was published by the corporate network Business & Society Belgium (2011) and is based on a survey among SMEs and large corporations. Business & Society Belgium (2011) emphasises that the idea of sustainable development has now been taken up by Belgian companies. However, in their more general analysis of sustainability management tools, the authors find that Belgian companies hesitate to actually implement these tools. But, at the same time, they recognise that improvements have been made.

These studies provide valuable insights into the Belgian approach to CSR and can thus be used as a basis for a more detailed analysis. Such an analysis seems necessary, since the existing investigations offer only limited details about corporate intentions, their degree of integrating sustainability into the core business and the actual implementation of specific measures.

Furthermore, most recent studies also address SMEs. These studies are thus unable to provide specific findings on the particular characteristics of large firms. In addition to addressing this deficit, this book chapter adds a truly international perspective, since the analysis is based on a survey which was carried out in a standardised manner in 11 countries coordinated by the Centre for Sustainability Management (CSM) in Lüneburg, Germany (see Schaltegger et al. 2013). While Louche et al. (2009) proposed making a preliminary European comparative study, none of the above studies compares Belgium with other countries.

However, an international comparison is of great importance, since the prevailing national business systems (i.e. the social, cultural, political and economic context) have a huge influence on the national understanding of corporate sustainability and the implementation of corresponding corporate strategies and measures (Doh and Guay 2006; Campbell 2007; Tempel and Walgenbach 2007; Matten and Moon 2008).

To investigate the contemporary state of the art in sustainability management and to benchmark Belgian sustainability management practices with the international state of the art, this chapter highlights best practices and identifies ways to improve the performance of Belgian companies by taking the specific Belgian context into account.

5.1.2 The Belgian Sample

A population of 138 large companies operating in Belgium was selected on the basis of two sources. First, the 100 largest Belgian companies (in terms of revenues) were identified using Bel-first, an official database (Bureau van Dijk 2011). Nine of these companies had to be subsequently excluded from the analysis because for instance their sustainability management was managed by a parent company. Ultimately 91 of these companies were provided with a questionnaire.

Second, as Business & Society Belgium is one of the most influential Belgian networks for sustainability management (Louche et al. 2009; Business & Society Belgium 2011), the 47 largest member companies of this network (revenues greater than €50 million) were also contacted and a questionnaire was sent to them.

Table 5.1 Number of employees ($n_{BEL} = 22$, $n_{INT} = 465$)

Number of employees	Number of sampled firms (Belgian sample)	Percentage (Belgian sample)	Percentage (Intel average)
≤50	0	0	0.4
51–250	3	13.6	2.8
251–1,000	6	27.3	9.7
1,001–10,000	8	36.4	40.9
10,001–100,000	5	22.7	38.7
>100,000	0	0	7.5

Table 5.2 Revenues ($n_{BEL} = 22$, $n_{INT} = 468$)

Revenues	Number of sampled firms (Belgian sample)	Percentage (Belgian sample)	Percentage (Intl average)
>€50 – €500 million	8	36.4	13.9
>€500 – €1,500 million	5	22.7	17.1
>€1,500 – €2,500 million	1	4.5	13.2
>€2,500 – €5,000 million	1	4.5	14.5
>€5,000 – €50,000 million	4	18.2	31.4
>€50,000 million	3	13.6	9.8

After several follow-up contacts, the questionnaire was completed by 22 of these 138 corporations, corresponding to a response rate of 15.9 %. A brief description of the specificities of the Belgian sample is given in Tables 5.1 and 5.2.

The Belgian sample is characterised by a set of *smaller companies* (in terms of employees and revenues) than the international average. Some of the sampled firms are almost medium-sized firms according to European Commission (2003) criteria.[1] Indeed, 13.6 % of the sampled Belgian firms employed less than 250 employees in 2012, while this category of firms only represents 3.2 % of the international average. In addition, while 40.9 % of the Belgian firms have less than 1,000 employees, these firms only represent 12.9 % of the international average. There are no firms with more than 100,000 employees in the Belgian sample, while these firms represent 7.5 % of the international average. Finally, 59.1 % of the sampled firms have revenues of up to €1,500 million, while this category represents only 31.0 % of the companies on international average.

Consistent with the description of the Belgian economic landscape, a broad distribution is observed when considering the share of non-domestic sales. Two extreme cases can be differentiated. On the one side, the sample is composed of firms that are very domestically oriented. Indeed, in 38.5 % of the sampled firms non-domestic sales represent less than 20 %. On the other side, the sample is composed of a number of very internationally oriented firms. In more than 30 % of the sampled firms, non-domestic sales represent more than 80 %.

---

[1]The EU defines a large company as one with a headcount of more than 250 people; turnover greater than €50 million; or a balance sheet total of more than €43 million.

Table 5.3 Industry ($n_{BEL} = 22$, $n_{INT} = 468$)

Industry	Number of sampled firms (Belgian sample)	Percentage (Belgian sample)	Percentage (International average)
Industry, capital goods and building	4	18.2	22.6
Consumer goods, trade and logistics	5	22.7	21.6
Finance and services	10	45.5	32.1
Commodities, auxiliary material, energy, chemical and pharmaceutical industry	3	13.6	23.7

More than 45 % of the sampled firms belong to the finance or services industries (see Table 5.3), which is consistent with their above-average share in the Belgian business landscape (SPF Economie, P.M.E., Classe Moyenne et Energie 2011).

5.2 Analysis

In most of the issues addressed by the International Corporate Sustainability Barometer (ICSB), we observe that Belgian firms are positioned below the international average. Different possible reasons for this observation will be developed in this section as well as in the next section with reference to the specific Belgian context (see Sect. 5.1.1).

5.2.1 Intention

5.2.1.1 Influence of Stakeholders

Belgian firms rate the influence of *internal stakeholders* below the international average. However, the most and least influential internal stakeholders are quite similar in Belgium and in the international average. The sustainability and the public relation/communication units as well as the top management are perceived as key drivers for sustainability implementation, while the finance and accounting/control units have the most restricted impact. These results are consistent with the findings made by Business & Society Belgium (2011).

The perceived influence of *external stakeholders* is also generally lower in Belgium than on international average (see Fig. 5.1). This observation is also consistent with the findings from previous national studies, since FEB (2007) and Louche et al. (2009) also reveal deficiencies in the stakeholder management of Belgian companies.

Firstly, Fig. 5.1 shows that the Belgian respondents particularly stress the influence of (1) NGOs, (2) the community, (3) investors, (4) the media and (5) international authorities (such as the United Nations).

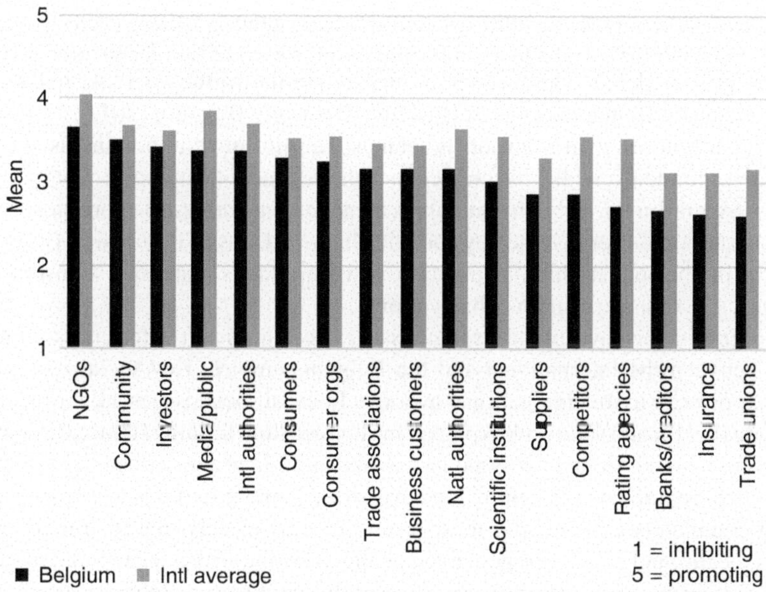

Fig. 5.1 Influence of external stakeholders (n_{BEL} ranging from 14 to 22, n_{INT} ranging from 393 to 450)

Secondly, consumers and customers as well as consumer organisations have a real influence, but their impact is more limited than the impact of the five categories of external stakeholders listed above.

Thirdly, competitors, suppliers, trade unions, banks and insurance companies are found to be the least influential external stakeholders. The key actors belonging to the companies' competitive environment thus have little influence in Belgium. When comparing the national average with the international average, we note that this observation is particularly true in Belgium. This shows that, even if more and more firms consider sustainability issues as a source of opportunity and innovation for new products and services (Business & Society Belgium 2011) and even if some signs indicate that Belgian firms perceive a relationship between sustainability and creating a competitive advantage (e.g. the influence of consumers and customers), they do not clearly and comprehensively link sustainability management to competitive advantage (e.g. with an opportunity to gain an advantage over their competitors). Finally, rating agencies are perceived as less influential in Belgium (one of the lowest scores).

5.2.1.2 Issues Managed

Even if they are less closely managed, the top sustainability issues managed by Belgian firms are similar to those observed on international average. Also, the survey shows they are generally aligned with stakeholder demands.

As far as *environmental issues* are concerned, Belgian companies closely manage energy consumption as well as emissions and waste, while biodiversity is less commonly considered. These observations are similar to the international average and consistent with the findings of Business & Society Belgium (2011).

It is worth noting that transport is seen as the third most important issue by the Belgian respondents, while this challenge is the second to last issue on international average. Accordingly, Belgian companies manage transport issues more closely than their international peers. Some key specific characteristics of the country (its central location in Europe, its high population density and its important transportation infrastructure) can explain this observation.

Concerning the investigated *social issues*, no significant differences could be found between the international and the Belgian samples. Key social issues have a similar priority to the international average. Internal social issues are more closely managed (health and safety, workplace/employment and training) than external ones (e.g. child labour). In fact, external issues like child labour or consumer protection are strongly regulated in Belgium. However, based on our findings, the legal context cannot completely explain why these issues are less closely managed than others. Indeed, health and safety are perceived as the most important issue to be managed by Belgian firms, although they are also highly regulated in Belgium. In addition, we observe that diversity issues have a great importance in Belgian companies. This can be explained by related academic activities and legislation. Business & Society Belgium (2011) also found an increasing corporate interest in diversity issues over the last 2 years.

5.2.1.3 Inhibiting Factors for Sustainability Management

The lack of financial capacities is considered as the most inhibiting factor for engaging in sustainability management in Belgium. This factor has also been highlighted as a top inhibiting factor in the international comparison. Nevertheless, the lack of personnel capacities has been emphasised more in the international average. The smaller size of the firms in the Belgian sample can explain why they place a greater emphasis on the lack of financial resources.

5.2.2 Integration

5.2.2.1 Connection with Core Business

Compared to the international average, Fig. 5.2 shows that a particularly high percentage of Belgian firms *connect sustainability to most segments* of their core business or even connect sustainability in a consistent manner *to all segments* of their core business.

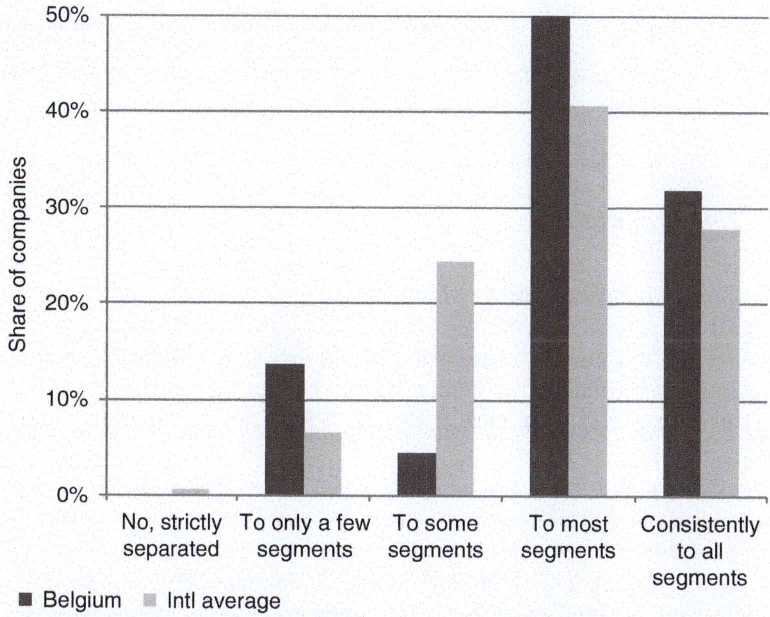

Fig. 5.2 Connection with core business ($n_{BEL} = 22$, $n_{INT} = 457$)

This observation shows that Belgian firms tend to connect sustainability issues to a wide range of segments of their core business more frequently than the international average and that they are thus particularly interested in finding opportunities to link sustainability with their core business. This observation is consistent with the findings from previous studies (Business & Society Belgium 2011) underlining that some supporting networks heavily promote the materiality principles.

5.2.2.2 Involvement of Other Departments

Similar to the international average and consistent with the findings of recent Belgian surveys, the CSR/sustainability department commonly holds the main responsibility for sustainability management. Other departments like the public relations/communication department or the top management are commonly involved. Consistent with previous literature (Zvezdov et al. 2010; Ballou et al. 2012), we observe that the finance and accounting/control departments are the organisational units least involved in sustainability management, although their expertise is needed (Zvezdov et al. 2010; Ballou et al. 2012).

This observation challenges somewhat the findings of the recent study carried out by Business & Society Belgium (2011:3), which underlines that "*sustainability tends to be integrated into the practices of all companies' departments, from the design of products and services to after-sales service*".

5.2.3 Implementation

5.2.3.1 Tools for Sustainability Management

As holds for most of the issues addressed in this survey, the average awareness and the application of sustainability management tools are below average in Belgium (average number of applied tools in Belgium: 17.95; international average: 27.17; see Appendix 5.1).

Particularly large differences can be observed for integrated sustainability management tools, which usually refer to the term "sustainability" (e.g. sustainability marketing, sustainability benchmarking). One reason for this might be that reference to the terms "social" or "environmental" is more common in Belgium. Nevertheless, these tools are of central importance for connecting sustainability with the core business, since they address all dimensions of sustainability simultaneously.

Similar negative deviations of the Belgian results can be observed for accounting and controlling tools (e.g. eco-budgeting, sustainability accounting, environmental accounting, ABC analysis, environmental cost accounting, material and energy flow accounting, social cost accounting, eco-investment accounting). Based on various documents and our own experience, we identify two major reasons for this observation. On the one hand, until now, the key supporting actors in Belgium (scientific community, networks, platforms and government authorities) have not extensively promoted these aspects and tools (via conferences, press articles, practitioner reports or specific workshops). On the other hand, these are technical tools that require some knowledge and expertise (in accounting, for example) that the respondents (sustainability managers) do not necessarily have. These tools require the support of additional actors (e.g. in the accounting department), who are usually not involved in corporate sustainability management (see Sect. 5.2.2).

Nevertheless, we observe a particularly high awareness of a limited number of specific sustainability management tools promoted by key supporting actors (e.g. social audits, sustainability supply chain management or stakeholder dialogue) or imposed by the Belgian legislation (e.g. social report).

5.2.3.2 International Standards

As shown in Fig. 5.3, the awareness and application of international standards is generally more limited in Belgium than in the international average. The fact that

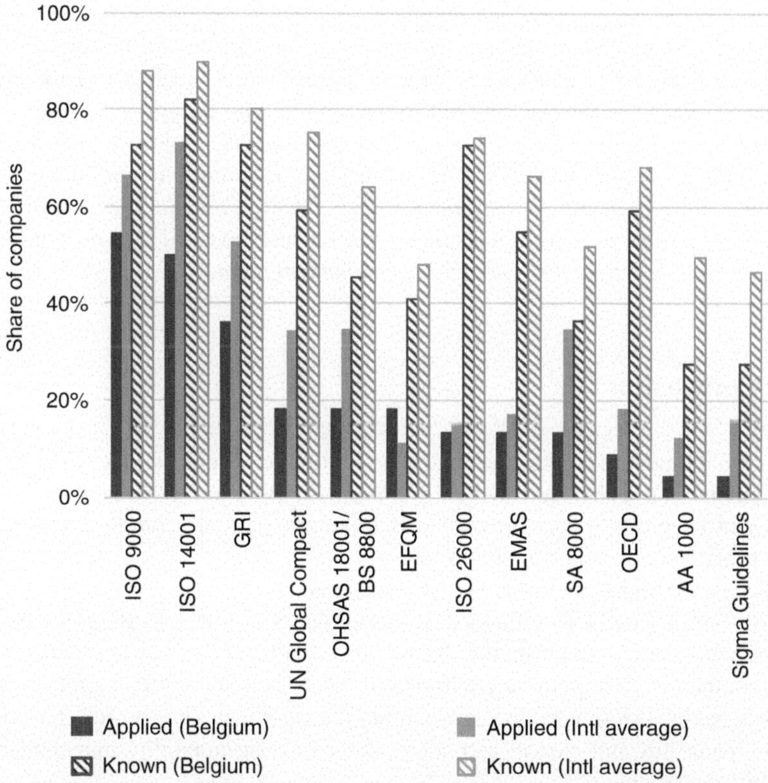

Fig. 5.3 Awareness and application of international standards ($n_{BEL} = 22$)

the Belgian sample is characterised by a set of *smaller companies* (in terms of employees and revenues) than the international average is a possible explanation for this observation.

Some exceptions have nevertheless been detected. Firstly, even if its application is still limited, the awareness and application of the ISO 26000 standard in Belgium is similar to the international average. Secondly, the SA 8000 and the EFQM standards are less well known, but they are applied more commonly in Belgium than on international average. One possible reason for this might be that key actors in Belgium (networks, platforms or scientific community) have particularly promoted these two international standards over the last years.

In Belgium, the most frequently known standards are ISO 14001, ISO 9000, ISO 26000 and the GRI guidelines. Consistent with the findings by Business & Society Belgium (2011), the most applied standards are ISO 14001, ISO 9000 and the GRI guidelines. Other standards like AA1000, OHSAS or Sigma guidelines are not known by a large proportion of the Belgian firms and are thus not frequently applied.

When considering international standards, it is worth mentioning that we observe a less clear link between awareness and application than we did for sustainability management tools. For example, standards like ISO 26000; EMAS or the OECD Guidelines are well known but not often applied in Belgium. Louche et al. (2009:135) provide one potential explanation for this observation. "*Companies are aware of the existing international standards and guidelines but a majority are not ready or willing to use/implement them mainly because they lack information, the firms do not perceive a direct relevance for their business and they do not always have the resources (time and financial) to implement them.*"

5.3 Conclusion

The International Corporate Sustainability Barometer confirms that sustainability management is a living and evolving concept in Belgium. However, as Louche et al. (2009) suggest, it also shows that Belgian practices are commonly below the international average and sometimes uneven.

This chapter highlights some key observations.

Firstly, the (perceived) influence of stakeholders is lower in Belgium than on international average. In particular, the perceived influence of key actors belonging to the companies' competitive environment (such as competitors, suppliers, trade unions or banks) is low as far as sustainability management is concerned. This shows that companies do not clearly and consistently link sustainability management to competitive advantage.

Secondly, even if they are less closely managed, top societal issues managed by Belgian firms are similar to those observed on international average. Nevertheless, transport is more closely managed in Belgium. The central location of Belgium in Europe, its high population density and its important transportation infrastructure can explain this observation.

Thirdly, a particularly high percentage of Belgian firms tend to connect sustainability to most segments of their core business or even connect sustainability consistently to all the segments of their core business. This observation shows that Belgian firms are particularly interested in finding opportunities to link sustainability with their core business.

Finally, the awareness and application of sustainability management tools as well as international standards are below average in Belgium. Especially integrative sustainability management tools (in contrast to environmental or social management tools) referring explicitly to the concept of "sustainability" as well as specific accounting and control tools are less frequently known and applied in Belgium.

These findings can be explained by several factors.

Firstly, previous research revealed that larger corporations tend to engage in sustainability issues more comprehensively. More specifically, they were found to have higher degrees of formalisation of their sustainability initiatives, to be more likely to introduce control mechanisms for sustainability and to show better overall social and environmental performances (Brammer and Pavelin 2006; Business & Society Belgium 2011; Gallo and Christensen 2011). As the average company size in the Belgian sample is smaller than that of the international sample, some of the negative deviations might be partly explicable by the effect of company size.

Secondly, this chapter highlighted the particular importance of *crucial supporting actors* (such as networks, platforms and scientific institutions) in the Belgian context. As sustainability management is a relatively young concept in Belgium, Belgian practices heavily depend on the knowledge these actors diffuse.

Thirdly, a large proportion of the sampled firms belong to foreign multinational corporations. They are thus the Belgian headquarters of these groups. We could imagine that, as the top management often drives sustainability, some of the managers in the Belgian subsidiaries are less able to disclose information about sustainability management practices and tools.

Fourthly, people from the finance and accounting/control areas are the least involved in sustainability management although their expertise could be very important to improve these aspects (Ballou et al. 2012).

Fifthly, sustainability management takes time (York 2009). To move towards sustainability, companies need to modify and, ideally, rethink the way they do business (new ways of purchasing, producing, distributing, communicating, etc.). As it is a relatively new concept in Belgium compared to some other countries (like Germany or the UK), it is quite unsurprising that Belgian firms have not made as much progress as some of their international counterparts.

Based on these observations, several paths for improve sustainability management practices in Belgium could be investigated.

Firstly, even if it has improved greatly over the last year, scientific institutions and academic research centres could engage more with businesses and diffuse knowledge about sustainability management tools.

Secondly, the promotion of sustainability management tools (especially accounting and control tools) by key supporting actors (like networks, platforms or government) is essential to improve sustainability management practices in Belgium.

Thirdly, the involvement of actors from the finance and accounting/control areas would also be crucial to improve sustainability management and measurement in Belgian firms (Zvezdov et al. 2010; Ballou et al. 2012). Indeed, some of the accounting management tools are very technical. Their application thus requires specific accounting expertise, which sustainability managers usually do not possess. To increase the involvement of actors belonging to the finance and accounting/control departments, and thus to enable the application of the respective tools, corporate collaborations with accounting and controlling organisations (such as the Belgian Institute for Accountants) could be a promising first step.

Appendix 5.1: Application of Sustainability Management Tools

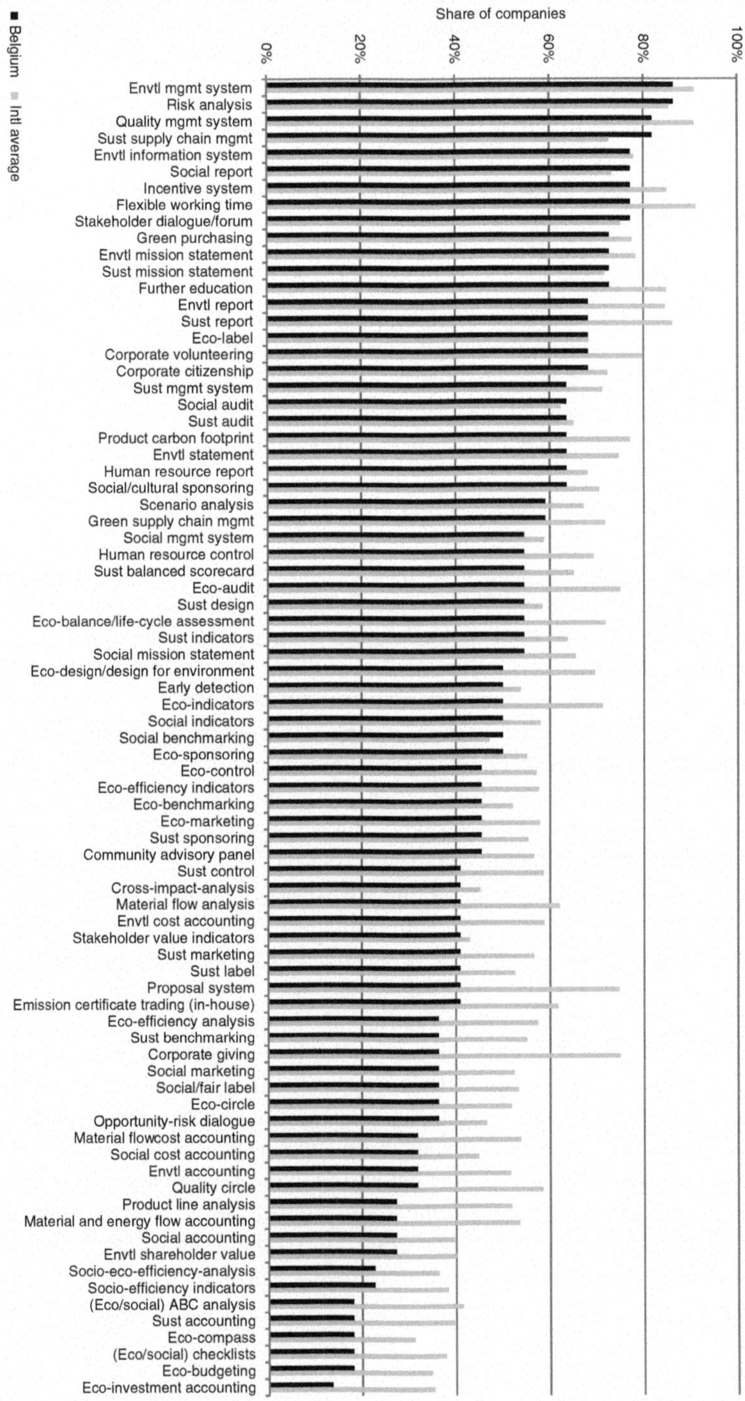

References

Ballou B, Casey R, Grenier J, Heitger D (2012) Exploring the strategic integration of sustainability initiatives: opportunities for accounting research. Account Horiz 26(2):265–288

Brammer S, Pavelin S (2006) Corporate reputation and social performance: the importance of fit. J Manag Stud 43(3):435–455

Bureau van Dijk (2011) Bel-first. Belgian and Luxembourgish financial company information and business intelligence. https://belfirst.bvdinfo.com/version-201393/Home.serv?product=belfirstneo. Accessed Mar 2012

Business & Society Belgium (2005) Le baromètre CSR. Business and Society Belgium Magazine, Business & Society Belgium, Brussels

Business & Society Belgium (2011) Le baromètre CSR. Business and Society Belgium Magazine, Business & Society Belgium, Brussels

Campbell J (2007) Why would corporations behave in socially responsible ways? An institutional theory of corporate social responsibility. Acad Manage Rev 32(3):946–967

Doh J, Guay T (2006) Corporate social responsibility, public policy, and NGO activism in Europe and the United States: an institutional-stakeholder perspective. J Manag Stud 43(1):47–73

European Commission (2003) Commission recommendation of 6 May 2003 concerning the definition of micro, small and medium-sized enterprises. Off J Eur Union 124:36–41

FEB (Federation of Enterprises in Belgium) (2007) RSE: une dynamique à renforcer. Plutôt qu'à affaiblir! Focus 17:10

Gallo P, Christensen L (2011) Firm size matters: an empirical investigation of organizational size and ownership on sustainability-related behaviors. Bus Soc 50(2):315–349

Keman H (2008) The low countries: confrontation and coalition in segmented societies. In: Colomer JM (ed) Comparative European politics. Routledge, London/New York, pp 208–245x

Louche C, van Liedekerke L, Everaert P, LeRoy D, Rossy A, d'Huart M (2009) Belgium. In: Idowu SO, Leal Filho W (eds) Global practices of corporate social responsibility. Springer, Berlin/Heidelberg, pp 125–147

Matten D, Moon J (2008) 'Implicit' and 'explicit' CSR: a conceptual framework for comparative understanding of corporate social responsibility. Acad Manage Rev 33(2):404–424

Mazijn B, Gouzee N (2007) Making sustainable development reality. 10 years of Belgian federal sustainable development strategy. Federal Public Planning Service, Brussels

O'Brien P, Carey D, Høj J, Woergoetter A (2001) Encouraging environmentally sustainable growth in Belgium. OECD Economics Department Working Paper, no 300. OECD, Paris

Schaltegger S, Harms D, Hörisch J, Windolph SE, Burritt R, Carter A, Truran S, Crutzen N, Ben Rhouma A, Csutora M, Tabi A, Kokubu K, Kitada H, Haider MB, Kim J-D, Lee K-H, Moneva JM, Ortas E, Álvarez-Etxeberria I, Daub C-H, Schmidt J, Herzig C, Morelli J (2013) International corporate sustainability barometer: a comparative study of 11 countries. Centre for Sustainability Management, Lüneburg

SPF Economie PME, Classe Moyenne et Energie (2011) Panorama de l'Economie Belge. Service public fédéral Economie/P.M.E./Classes moyennes et Energie, Brussels

Tempel A, Walgenbach P (2007) Global standardization of organizational forms and management practices: what new institutionalism and the business-systems approach can learn from each other. J Manag Stud 44(1):1–24

York J (2009) Pragmatic sustainability: translating environmental ethics into competitive advantage. J Bus Ethics 85(1):97–109

Zvezdov D, Schaltegger S, Bennett M (2010) The increasing involvement of accountants in corporate sustainability management. J Asia-Pacific Cent Environ Account 16(4):20–31

Chapter 6
International Corporate Sustainability Barometer 2012: Sustainability Management in France

Amel Ben Rhouma, Claude Francoeur, and Guillaume Robin

Abstract This chapter summarises the results for France as part of the International Corporate Sustainability Barometer (ICSB). The introduction and the first section describe the context of sustainability in France and the characteristics of the French companies surveyed. The second section presents the main findings and some comparisons between our sample of French companies and the international average regarding the three main topics of the International Corporate Sustainability Barometer. Concerning the intention to manage sustainability issues, French firms are found to be most strongly influenced by national authorities. With regard to the integration of sustainability into the organisation, the results show that in France social and environmental issues affect the majority of organisational units to a lesser degree than is the case on international average. In addition, differences exist between organisational units in France. More specifically, sustainability-related and externally-oriented units, such as CSR and communication, are most supportive of sustainability. By contrast, performance-oriented units, such as finance and accounting, are less supportive. Lastly, the actual implementation of corporate sustainability is characterised by the fact that the application of sustainability management tools is less common among French companies when compared to the international average. Nevertheless, French companies frequently measure the impact of their sustainability measures.

A. Ben Rhouma (✉)
CEDAG Management EA-1516 IUT, Department of Management and Administration, University Paris-Descartes, Paris, France
e-mail: amel.ben-rhouma@parisdescartes.fr

C. Francoeur
Department of Accounting Studies, HEC Montreal, Montreal, Canada
e-mail: claude.francoeur@hec.ca

G. Robin
IUT – Department of Marketing Techniques, University Paris-Descartes, Paris, France
e-mail: guillaume.robin@parisdescartes.fr

S. Schaltegger et al. (eds.), *Corporate Sustainability in International Comparison*, Eco-Efficiency in Industry and Science 31, DOI 10.1007/978-3-319-06227-3_6,

© Springer International Publishing Switzerland 2014

6.1 Introduction

6.1.1 The French Context

In France, there is an ongoing debate on a proper translation of the English concept of sustainable development. Should it be *développement soutenable* (which means 'sustainable') or *développement durable* (which means 'durable')? The same debate concerns the translation of the concept sustainability management, with some preferring *management responsible* instead of *management durable*. According to Le Bas (2004), responsible management is another term for corporate social responsibility (CSR),[1] which was introduced in France through legal and political initiatives. In the social dimension, the decree of March 23, 1967 Article 148–2 requires a company to disclose information about the social consequences of its activities in its annual report (such as the number of employees, working conditions and salary policies).

Law No. 77–769 of 12 July 1977 requires French companies with more than 300 employees to collect and publish an extensive social balance sheet (*bilan social*). The concepts of 'added social value' and 'human capital' have become more and more prominent in management practices. The management of risk is regulated by the Act of 31 December 1991, which requires companies to promote the health and safety of employees in the workplace. According to Bourion (2006), these efforts are attempts to revalue the social dimension of companies but they are primarily focused on increasing financial efficiency.

In the political field, since the Delors plan of 1993, initiated by the French president of the European commission Jacques Delors, called for a social dimension to be added to European policy, the number of initiatives has increased. For instance, the Lisbon European Council of 2000 set strategic goals to reach sustainable growth and quantitative and qualitative improvements in employment and social cohesion. In 2001, the European Union published a green paper to extend the discussion on CSR. Since this publication, CSR has developed into a key subject for all publicly listed firms and political measures to promote CSR in France have sharply increased. These measures include:

- Article 116 of Law No. 2000–420 of 15 May 2001 on New Economic Regulations (called the NRE Act), which requires French companies to include social and environmental information in their annual reports. This law applies to 700 French listed companies.

[1]During a meeting with 18 directors of SD/CSR at large French companies, representatives were asked to describe 'durable management' and 'responsible management'. Of the 18 directors, 6 stated there is a difference between the two terms, 8 understood the terms to be interchangeable, 3 preferred using only the term 'responsible management', and just 1 preferred using the term 'durable management'.

- The 19 February 2001 Fabius Act on employee savings schemes and the Act of 17 July 2001 on pension reserve funds, which both promote socially responsible investment. These regulations explain the social, environmental or ethical issues that an asset management company is supposed to address when purchasing or selling securities. These laws require these firms to justify that their investment policy is socially responsible.
- The 30 July 2003 law on the control of major-accident hazards involving dangerous substances, which reinforces the obligation of companies with one or more 'Seveso' sites to release environmental reports.

The legal environment of CSR in France has been completed by the 9 May 2001 Génisson Act and the 23 March 2006 Act on pay equity between men and women. The Génisson Act requires companies with more than 50 employees to establish equality in the workplace by specifying their indicators and including them in a comparative report. More recently, Article 225 of the Grenelle 2 Act extends the scope of the NRE Act to include all listed and non-listed French companies.

Beyond laws and regulations, French authorities have adopted a certification process for companies or products with the aim of promoting responsible management behaviour. In 2004, a law governing "Equality Label" certification was introduced to combat discrimination against women; in 2008, a "Diversity Label" certification to prevent social exclusion was also introduced.

The chapter is structured as follows. Section 6.1.2 describes the characteristics of the French companies surveyed. The second section presents the main findings and some comparisons between our sample of French companies and the international average regarding three main topics:

- Intention: why do French companies manage sustainability?
- Integration: how well are French companies able to integrate social and environmental policies into their core business and profit-making activities across all organisational units?
- Implementation: how systematically are sustainability management tools known and applied? How precisely is success in corporate sustainability measured and how intensively are relevant stakeholders involved?

6.1.2 Description of the French Companies Surveyed

The survey, created by the Centre for Sustainability Management at the Leuphana University Lüneburg (Schaltegger et al. 2013), was addressed to sustainability managers or CSR managers at French companies in the SBF 120 index.[2] Contact was made by phone, mail, or professional networks such as *Viadeo*, *LinkedIn* and the

[2]The SBF (Société Française des Bourses) 120 is a French stock index made up of the most important and actively traded publicly listed French firms, continuously quoted in the Primary Market or the Second Market (unlisted securities market). This index is more diversified than the Paris stock index CAC 40.

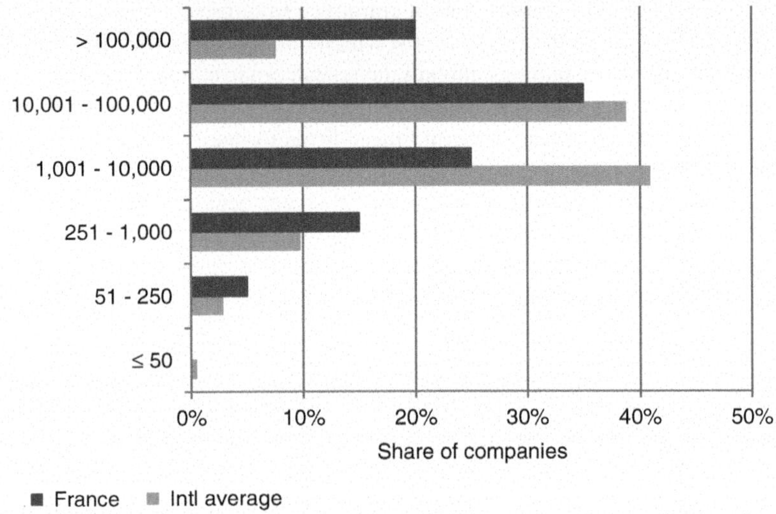

Fig. 6.1 Number of employees ($n_{FRA} = 20$, $n_{INT} = 465$)

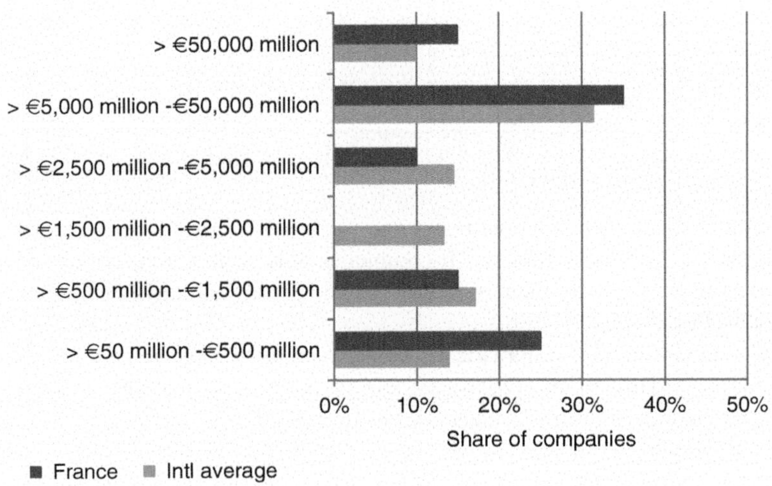

Fig. 6.2 Annual revenue ($n_{FRA} = 20$, $n_{INT} = 468$)

Circle of Sustainability/CSR Managers.[3] In total 93 questionnaires were sent out, which yielded 20 completed and returned responses. The response rate was 21.5 % and thus meets the validity requirements set by Bartlett et al. (2001).

This survey focuses on the largest French companies by revenue, taking all business sectors into account. Figures 6.1, 6.2 and 6.3 compare the characteristics of the French firms in the sample with the international average.

[3]This physical social network of sustainability and CSR managers was founded during the years 2000/2001 to develop informal exchanges between sustainability managers at French companies.

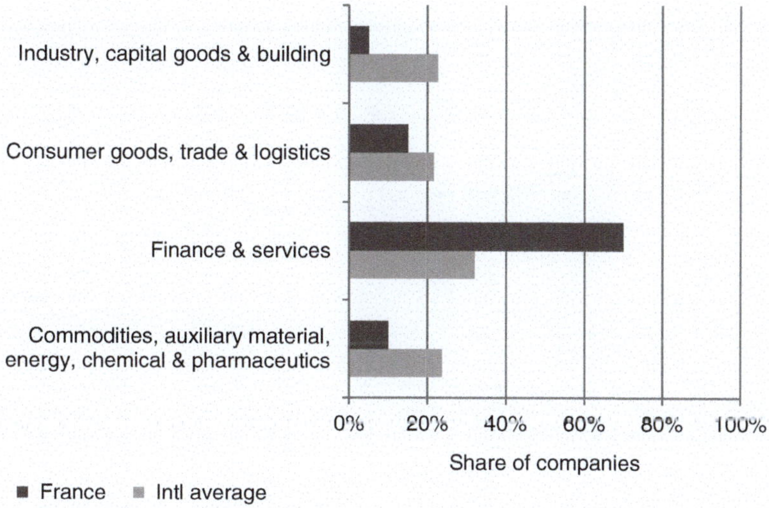

Fig. 6.3 Core business ($n_{FRA} = 20$, $n_{INT} = 468$)

Most companies employ more than 1,000 employees and have a significant effect on the labour market and on the country's economic development. The majority of the French companies surveyed (70 %) belong to the finance & services sector. Finance & service firms also make up the largest sector in the international average (32.1 %), although not an absolute majority. This difference is significant, and the French sample has by far the largest share of finance/service companies in our study.

6.2 Comparison of the French Results to the International Average

6.2.1 Intention

This section summarises the motivations that lead French companies to adopt a sustainability strategy, based on an analysis of the influence of external stakeholders and the management of sustainability issues.

6.2.1.1 Impact of External Stakeholders on the Integration of Sustainability

In most cases, external stakeholders have a neutral or positive impact on the integration of sustainability practices, both in France and in other countries (see Fig. 6.4). The following differences can be highlighted:

- Stakeholders with less impact in France relative to the international average are consumers, competitors, investors, banks and creditors, insurance companies, community, media and public, consumer organisations and trade unions.

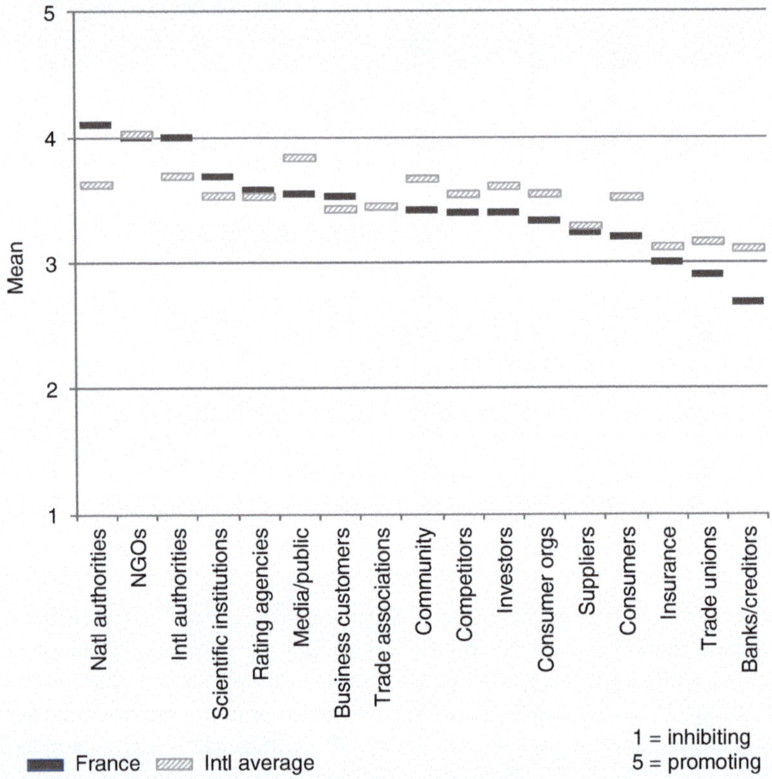

Fig. 6.4 Impact of external stakeholders on corporate sustainability (n_{FRA} ranging from 15 to 20, n_{INT} ranging from 393 to 450)

- Stakeholders that have a roughly equivalent impact in France and at the international level are rating agencies, NGOs, suppliers and trade associations.
- Stakeholders that have a stronger impact in France compared to the international average are business customers, national and international authorities as well as scientific institutions.

Regarding the factors promoting the implementation of sustainability, it appears that legislation (statutory rules and guidelines) and government incentives (subsidies, tax advantages, etc.) are viewed as more strongly promoting in France than within the international sample (see Fig. 6.5). Nevertheless, the know-how in the company, the self-commitment by business, the corporate philosophy, the commitment by the top management and the practicability of sustainability management methods are considered less promoting in France than in the international average.

Lack of government incentives, lack of personnel and financial capacities as well as lack of support by top management (among others) are assessed as factors more inhibiting in France than in the international average (see Fig. 6.6).

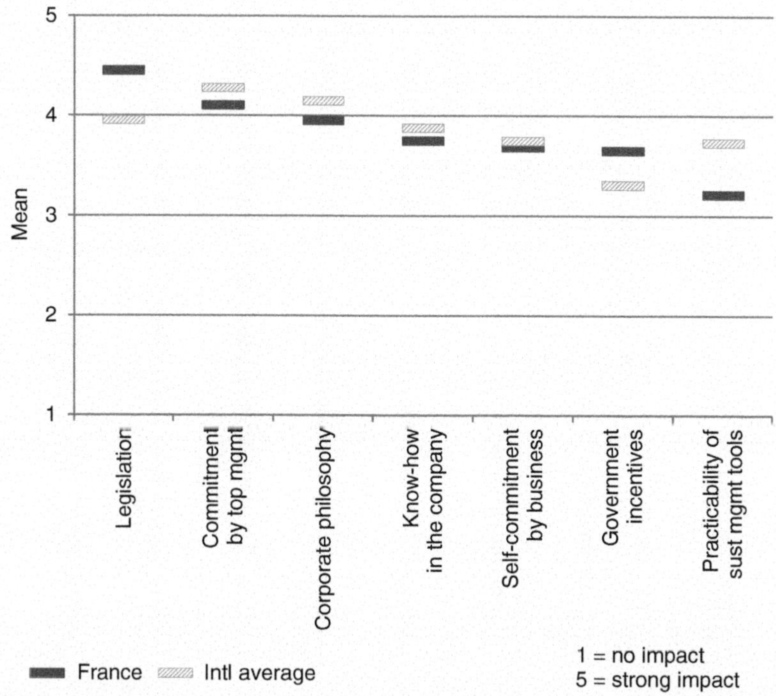

Fig. 6.5 Promoting factors for sustainability (n_{FRA} ranging from 19 to 20, n_{INT} ranging from 446 to 460)

6.2.1.2 Management of Sustainability Issues

Figure 6.7 shows that environmental issues are generally less managed in France than in the international average.

For social issues, great similarities exist concerning training and development as well as diversity and equal opportunity. However, differences exist in the management of other social issues. Workplace and employment as well as occupational health and safety are managed more closely on international average than in France. At the same time, issues such as consumer protection or child labour, forced and compulsory labour as well as freedom of association are all viewed as more closely managed by French companies.

The fact that environmental issues are assessed as less closely managed by French corporations seems in line with stakeholder demands. Figure 6.8 highlights that in France environmental issues are among the lowest stakeholder demands, and in particular engagement for biodiversity has the lowest priority. However, stakeholders in French companies would like stricter management of social issues. They show high values for the issues of training and development, diversity and equal opportunity, consumer protection, child labour, forced and compulsory labour as well as freedom of association.

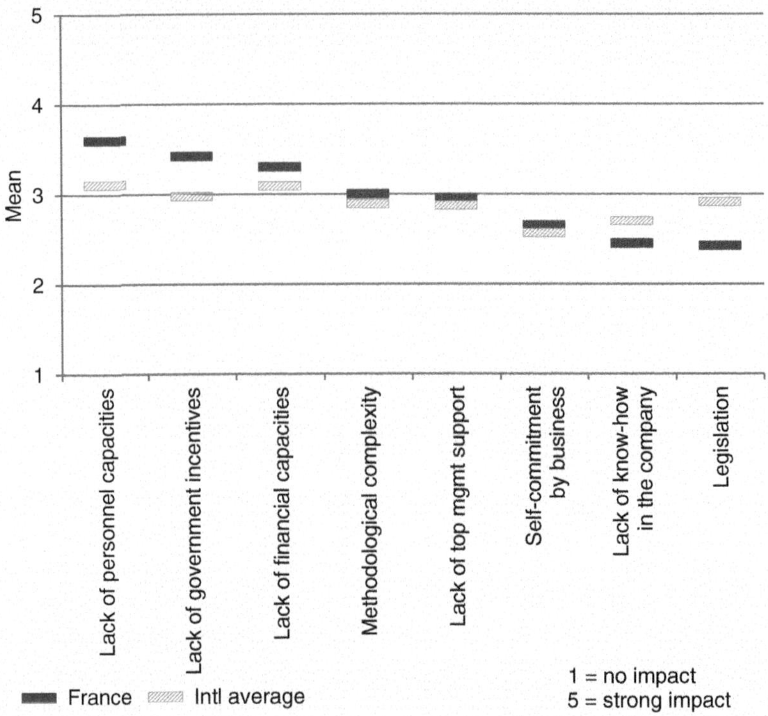

Fig. 6.6 Inhibiting factors for sustainability (n_{FRA} ranging from 19 to 20, n_{INT} ranging from 438 to 449)

6.2.1.3 Discussion and Implications of the Intention of French Companies to Manage Sustainability

According to these results, in France market-oriented stakeholders such as suppliers, insurance companies and banks are frequently ranked as less promoting sustainability than in the international average. However, the existing literature suggests that market forces can serve as a relevant trigger for corporate sustainability (Schaltegger and Lüdeke-Freund 2013). For example, Orlitzky et al. (2003) argue that managers who follow a sustainability approach by trying to satisfy their different stakeholder interests increase the company's adaptability to external demands. In addition, a company that reorients its research and development policy towards cleaner technologies and changes its production patterns is positioned before its main (non-socially responsible) competitors in the growing market for less polluting and consuming production using non-renewable resources. This is for example the case for STMicroelectronics, which estimates that resource savings (water, energy, chemicals) have reduced costs by 60 million euros over the period between 1994 and 2001 (Boasson and Wilson 2002). These potentials do not seem to be fully realised by large firms in France.

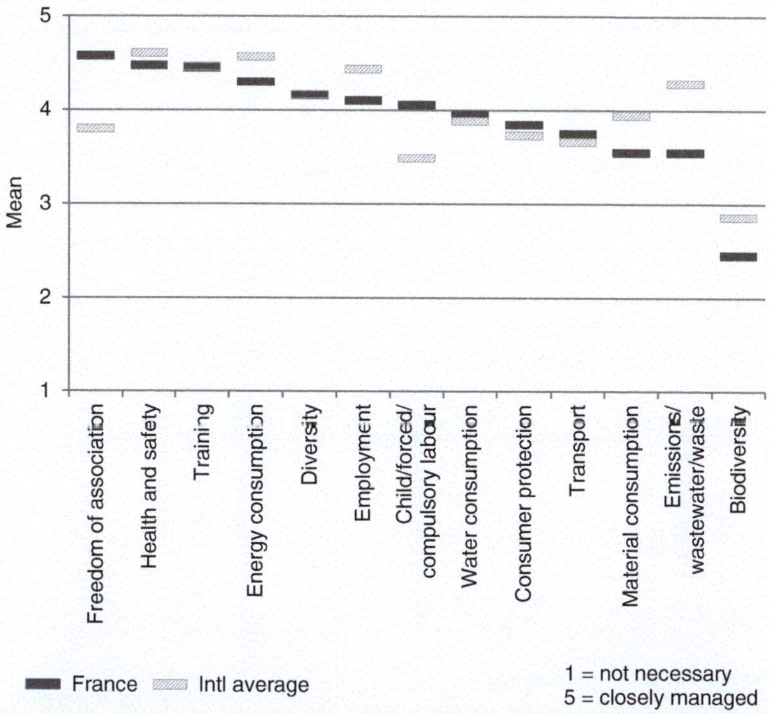

Fig. 6.7 Managed environmental and social issues (n_{FRA} ranging from 19 to 20, n_{INT} ranging from 442 to 463)

On the other hand, societal stakeholders seem to influence a French company's intention to address sustainability. This confirms the idea that "the enterprise is a social institution created by the society to which it is accountable and who is able to withdraw its privileges if it proves inadequate" (Gendron 2000:322). A major reason for the disjunction between private and collective interest is the presence of externalities. Accordingly, sustainability management can be seen as a way to internalise externalities, especially the negative externalities associated with pollution. It is also an alternative to coercive regulations, which are perceived by companies as generating additional costs (for negative effects on the production costs of pollution regulations, see Gray 1987 and Robinson 1995).

However, the intention of French firms to address sustainability is most influenced by national authorities. Legislation is therefore currently the most effective factor contributing to a firm adopting a sustainable strategy. This result can be explained by the importance of institutional isomorphism as defined by DiMaggio and Powell (1983) in encouraging managers to adopt a sustainability strategy. In the French case, the increase in shares held by foreign institutional investors made possible by financial deregulation, the succession of reports or governance codes

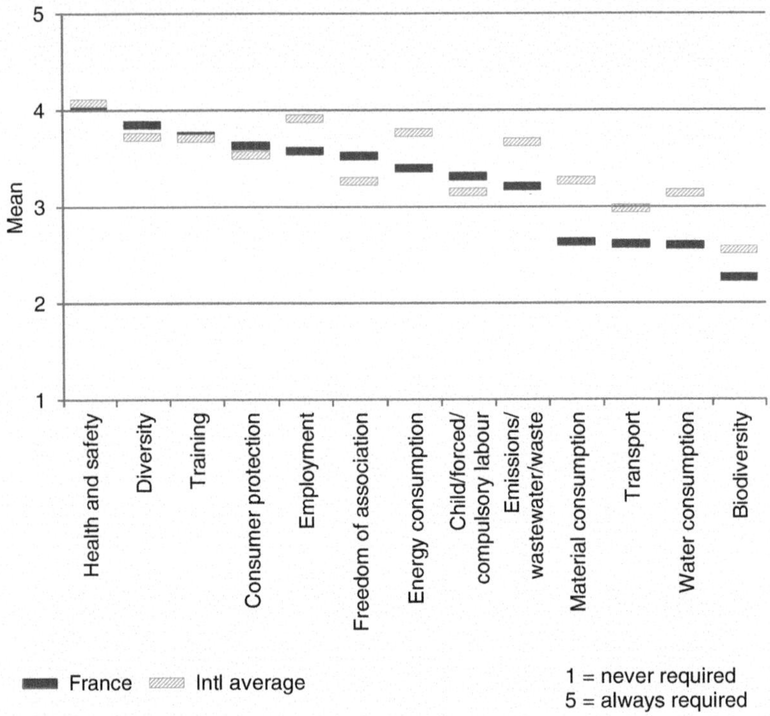

Fig. 6.8 Stakeholder demands to manage environmental and social issues (n_{FRA} ranging from 18 to 20, n_{INT} ranging from 443 to 461)

drawn up by employer associations[4] (Vienot I in 1995, Vienot II in 1999, Bouton 2002, the AFEP/MEDEF report 2008 and 2010),[5] the support of major international organisations such as the OECD and finally the modifications of the legal framework (especially the Law on New Economic Regulations in 2001, the Financial Security Law in 2003, the Grenelle 2 Law in 2010) are the main elements of a very rapid diffusion of sustainability initiatives. In only a short period of time, many French listed companies (including almost all CAC40 index) have been able to comply with good governance standards.

As demonstrated by the rapid development of the legal framework governing the activities of listed companies and stock exchanges and pressures from institutional investors, coercion is the main force of homogenisation of organisational sustainability practices in France. The other two forces of homogenisation, as described by DiMaggio and Powell, can also be found: mimicry, via the tendency

[4]These associations are AFEP, French Association of Large Companies (http://www.afep.com) and MEDEF (Movement of the Enterprises of France), the largest union of employers in France.

[5]These are the principal French reports and codes on corporate governance.

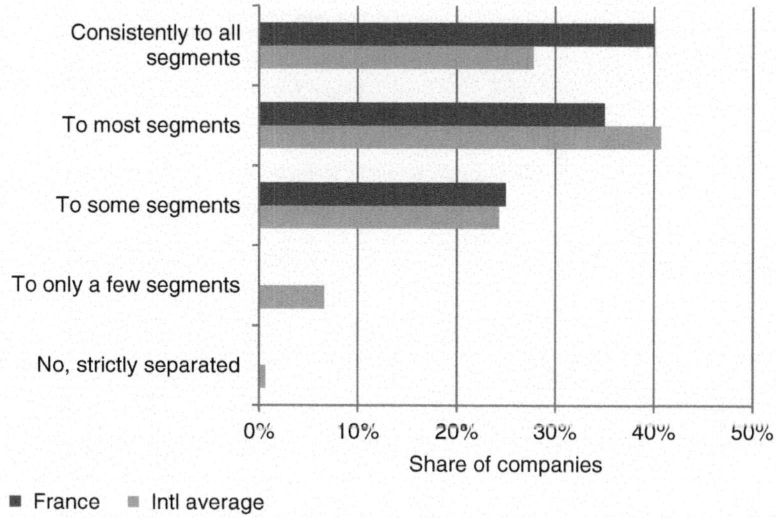

Fig. 6.9 Linking sustainability with the core business ($n_{FRA} = 20$, $n_{INT} = 457$)

of leaders to model their business strategy on what is perceived as a successful or 'normal strategy', and finally normative dynamics, which are conveyed by various organisations such as the OECD, European Commission, associations or consultancies (such as KPMG), the latter rating companies and countries based on their level of compliance with codified principles of corporate governance.

6.2.2 *Integration*

How well are French companies able to integrate social and environmental policies into their core business and profit-making activities across all organisational units?

6.2.2.1 Connection to Core Business

As shown in Fig. 6.9, about 75 % of French companies claim to link sustainability to most or all segments of their core business. On international average, about 68.5 % of all companies claim to link sustainability to most or all segments of their core activity.

Additionally, French companies were asked to provide examples of how they link their sustainability practices with their core business. Examples mentioned by the companies in the four sectors include:

- Industry, capital goods, building: consulting in energy and energy diagnosis, combination of gas and solar energy, rainwater recycling, energy performance, logistic impact reduction.

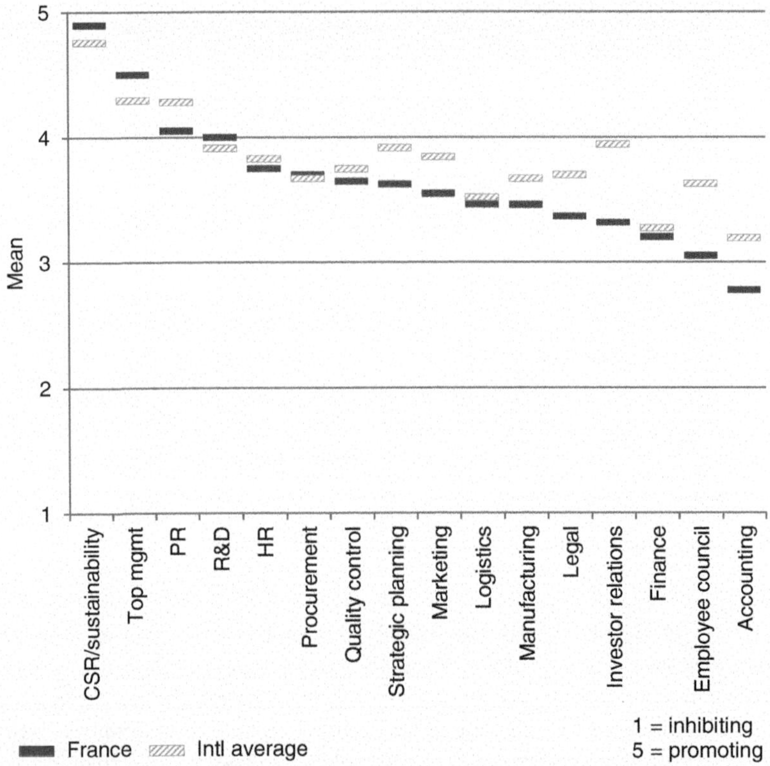

Fig. 6.10 Involvement of organisational units in corporate sustainability (n_{FRA} ranging from 13 to 20, n_{INT} ranging from 325 to 460)

- Consumer goods, trade and logistics: green products and services.
- Finances and services: more sustainable communications, more affordable and more accessible funding through supporting innovation and research and development.
- Commodities, auxiliary material, and energy, chemical and pharmaceutical industry: creating budgetary support to reduce externalities generated by business.

6.2.2.2 Involvement of Organisational Units

Another aspect of integration is the involvement of different organisational units in sustainability (Schaltegger et al. 2014). The results in Fig. 6.10 show that almost all organisational units promote sustainability practices or are at least neutral. There are several differences between France and the international average:

- Top management is seen as more promoting of sustainability in France than in the international average.

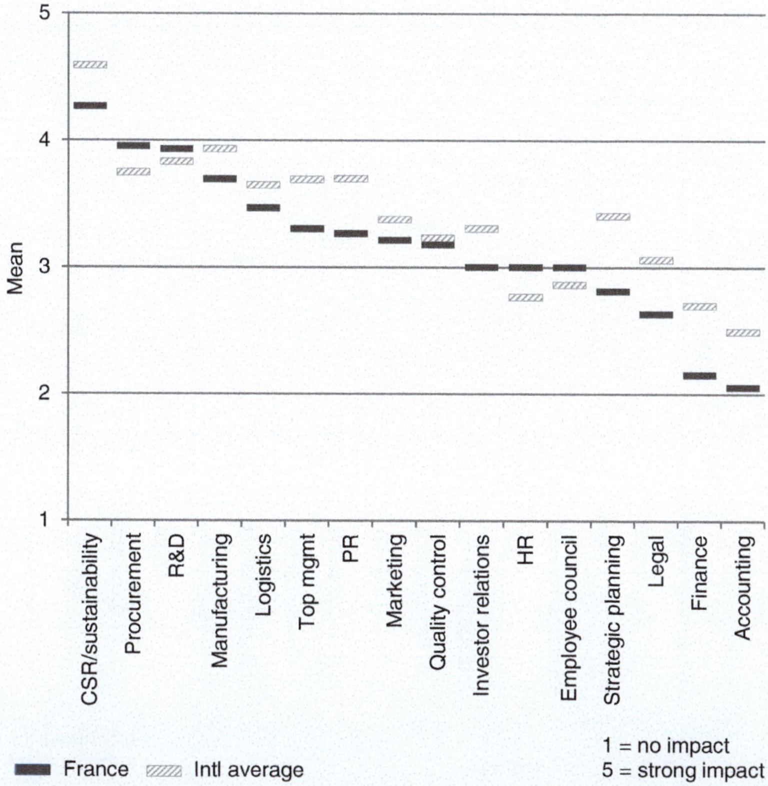

Fig. 6.11 Impact of environmental issues on organisational units (n_{FRA} ranging from 13 to 20, n_{INT} ranging from 329 to 457)

- In contrast, quality control, marketing, investor relations, human resources, legal department, finance and accounting, strategic planning, the employee council, PR and manufacturing are less supportive of sustainability practices in France.

It is important to note that the CSR/sustainability department, which receives the highest values in both samples, is found to be more supportive in France than in the international average. In France, finance and accounting departments are evaluated as neutral or less involved in sustainability than in the international average (at the international level, the finance and accounting department is viewed as being at least marginally supportive of sustainability).

In addition to the involvement of organisational units, the second aspect analysed is the impact of sustainability issues across organisational units (see Figs. 6.11 and 6.12).

In France, the results reveal large differences regarding the impact that environmental issues have on different organisational units. Following the CSR department, the most affected units are procurement and R&D. These two units, together with

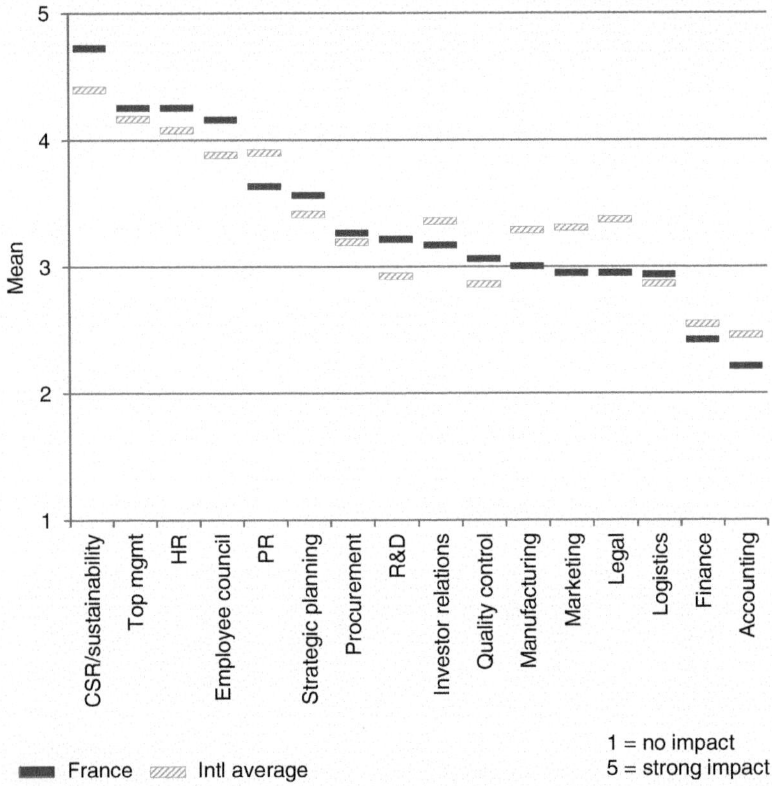

Fig. 6.12 Impact of social/societal issues on organisational units (n_{FRA} ranging from 13 to 20, n_{INT} ranging from 327 to 457)

HR and the employee council, are also more strongly affected in France than they are internationally.

With the exception of these last four organisational units, it seems that in France environmental issues have less impact on organisational units. Environmental issues are given little weight by the legal department as well as finance and accounting.

Looking at the impact that social issues have on the different business functions of a company, a common trend can be seen in France and in the international average, though some differences are worth noting (see Fig. 6.12):

- In general, the impact of social issues for some organisational units like top management, human resources and the sustainability departments is stronger than the impact of environmental issues.
- Similarly, in France, procurement and purchasing, R&D, logistics, quality control, strategic planning departments and the employee council are more affected by social issues than in the international average.

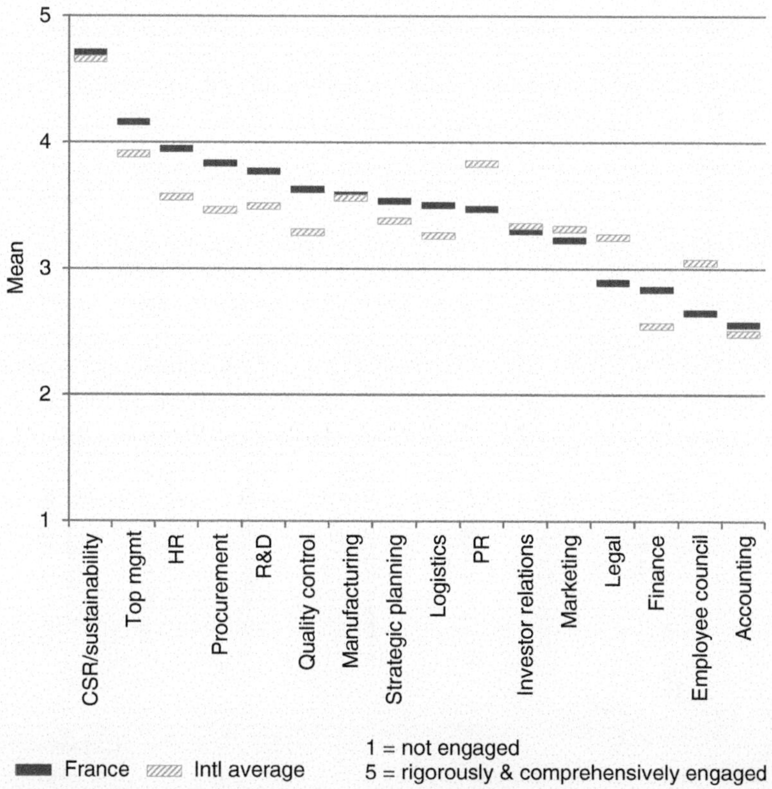

Fig. 6.13 Engagement of organisational units in sustainability measures (n_{FRA} ranging from 12 to 19, n_{INT} ranging from 286 to 418)

- However, the manufacturing, marketing, communication, investor relations, legal, finance and accounting departments are less affected in France relative to the international average.

The findings show that most organisational units are engaged in sustainability measures in France and in the international sample (see Fig. 6.13). Nevertheless, some differences can be highlighted. In France, particularly HR, procurement/purchasing and quality control are evaluated as more engaged in sustainability measures than on the international level, whereas the employee council, the public relations and legal department seem less engaged than on international average.

6.2.2.3 Drivers for Sustainability

In general, the implementation of sustainability measures within a company (see Fig. 6.14) is roughly the same in France as it is on international average. Several

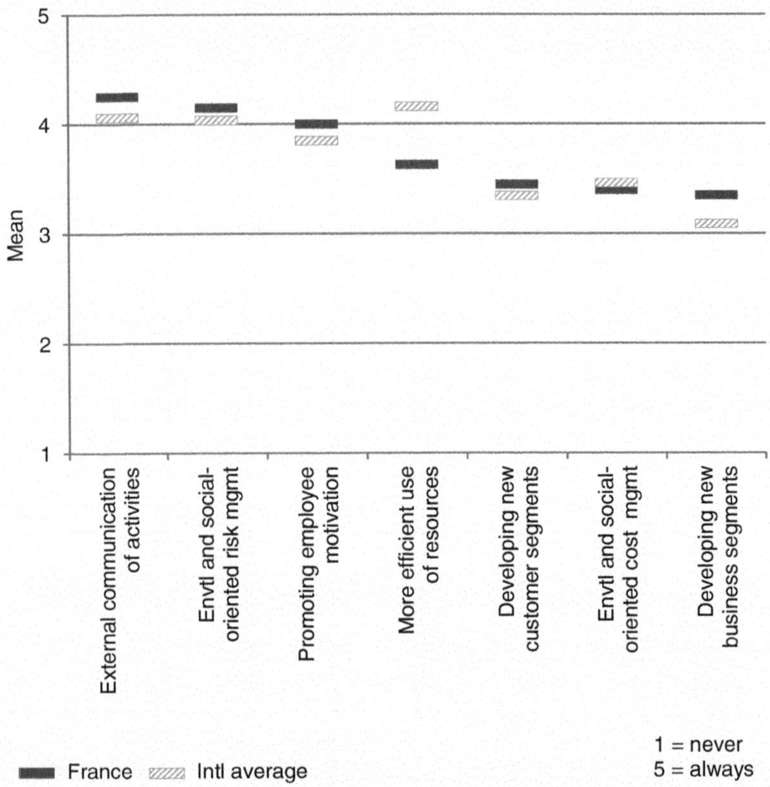

Fig. 6.14 Implementation of sustainability measures (n_{FRA} ranging from 19 to 20, n_{INT} ranging from 397 to 405)

important differences exist however. The development of new business segments related to sustainability, as well as the promotion of employee motivation and external communication of environmental and social activities are more widely implemented in France than they are internationally. By contrast, French companies are less interested in more efficiently using resources than on international average.

6.2.2.4 Discussion and Implications of the Integration of Sustainability by French Companies

In order to integrate sustainability issues into a company's economic activities, sustainability management should focus on those environmental and social issues that are linked to their value chain (Porter and Kramer 2006). According to our results, although a large share of French companies have made announcements concerning sustainability management, no company fulfils this commitment for all segments of their business at the same time. Nevertheless, as a result of Article

225 of the Grenelle 2 Law, which institutes a new process for sustainability reporting, we should expect future improvements in bundling all activities into an integrated sustainability strategy. This law focuses in particular on the need for companies to define their specific requirements regarding sustainability information and building material indicators. According to AA1000 AccountAbility Principles for Sustainable Development (1999), the materiality principle provides a means to define the most relevant and significant challenges for an organisation and its stakeholders. A significant material challenge is defined as a challenge that will influence the decisions, actions and performance of all units of an organisation as well as its core business.

Our results show that in France, social and environmental issues affect the majority of organisational units to a lesser degree than is seen in the international average. In addition, differences exist between organisational units in France. More specifically, sustainability-related and externally oriented units, such as CSR and communication, are most supportive of sustainability. By contrast, performance-oriented units, such as finance and accounting, are less supportive.

Lastly, the surveyed French companies are highly concerned with improving their reputation and maintaining their legitimacy in the eyes of their stakeholders (including their employees). They are, however, less concerned with using resources more efficiently.

Sustainability management should involve all management activities and departments. It must be more than the simple philanthropic activities managers undertake on behalf of their company. In brief, a business case must be made for sustainability.

6.2.3 Implementation

How systematically are sustainability management tools known and applied? How precisely is the success of corporate sustainability measured and how closely are relevant stakeholders involved?

6.2.3.1 Stakeholder Management

Table 6.1 below contains the ratings of both used and unused approaches supporting the interaction between companies and their stakeholders for the implementation of sustainability. French companies are close to the international average for most approaches. Notably, in France, approaches like observing stakeholders and informing stakeholders are used in most cases (not only on a case-by-case basis) by a majority of companies, and used more frequently than in the international average. In addition, involvement and consideration in decision-making processes and cooperating/networking to develop joint solutions are mostly used in specific cases in France and in the international sample. Furthermore, empowerment is less used in France than internationally.

Table 6.1 Management of stakeholder relations (per cent) (n_{FRA} ranging from 19 to 20, n_{INT} ranging from 438 to 458)

	France			International average		
	Not used	Used in specific case	Used in most cases	Not used	Used in specific case	Used in most cases
Observing stakeholders	0.0	40.0	60.0	8.1	48.4	43.5
Informing stakeholders (e.g., through websites, the press, etc.)	0.0	20.0	80.0	1.5	21.4	77.1
Dialogue with stakeholders/seeking advice (e.g., questionnaires, dialogue forums)	5.0	50.0	45.0	10.1	53.8	36.0
Involvement, consideration in decision-making process (e.g., in advisory boards)	20.0	65.0	15.0	27.5	53.4	19.1
Cooperating, networking to develop joint solutions (e.g., in working groups)	15.8	57.9	26.3	16.1	61.1	22.7
Empowerment (e.g., providing financial support)	55.0	35.0	10.0	43.0	49.5	7.4
Delegating decision-making authority	63.2	36.8	0.0	62.9	32.9	4.6

Table 6.2 Top 10 known sustainability management tools (per cent)

	France	International average
Flexible working time	80.0	91.2
Environmental management system	75.0	90.8
Quality management system	75.0	90.8
Sustainability report	80.0	86.1
Risk analysis	85.0	85.4
Incentive system	75.0	85.0
Further education	75.0	84.8
Environmental report	75.0	84.6
Corporate volunteering	65.0	79.7
Environmental mission statement	30.0	78.4

6.2.3.2 Sustainability Management Tools

Tables 6.2 and 6.3 contain the percentages of awareness and of the application of the respective top 10 sustainability management tools internationally.

Table 6.3 Top 10 applied sustainability management tools (per cent; $n_{FRA} = 20$)

	France	International average
Flexible working time	55.0	79.0
Environmental management system	55.0	77.6
Further education	70.0	76.7
Quality management system	50.0	75.6
Risk analysis	75.0	70.4
Incentive system	50.0	67.9
Sustainability report	70.0	63.2
Corporate giving	65.0	61.2
Corporate volunteering	50.0	59.5
Environmental mission statement	15.0	58.5

Table 6.4 Sustainability management standards guidelines applied (per cent; $n_{FRA} = 20$)

	France	International average
ISO 14001	60.0	73.1
ISO 9000	70.0	66.7
Global Reporting Initiative	65.0	52.8
OHSAS 18001/BS 8800	50.0	34.6
UN Global Compact	60.0	34.4
OECD	40.0	18.4
EMAS	15.0	17.1
Sigma Guidelines	45.0	16.0
ISO 26000	40.0	15.4
AA1000	20.0	12.2
EFQM	25.0	11.1
SA 8000	20.0	9.8

Results show that these tools are mostly less known and especially less applied in France than they are internationally. Risk analysis is the most commonly known tool and the most applied in France. Conversely, the environmental mission statement is by far the least recognised and applied tool in France.

Table 6.4 reports the percentages of application of sustainability management standards and guidelines. In most cases, these standards are applied more in France than they are internationally. Specifically, ISO 9000, Global Reporting Initiative, Global Compact, OHSAS 18001/BS 8800, Sigma Guidelines, ISO 26000, EFQM, AA1000, SA8000 and OECD Guidelines are applied more in France.

6.2.3.3 Measurement

Regarding the measurement of sustainability issues' impact, Table 6.5 shows that energy consumption is the environmental issue most frequently measured by the French companies surveyed. Training and development is the most frequently measured social issue in France.

Table 6.5 Measurements of sustainability issues' impact (per cent) (n_{FRA} ranging from 17 to 19, n_{INT} ranging from 425 to 454)

Sustainability issue	France	International average
Energy consumption	100.0	95.4
Water consumption	88.9	87.2
Material consumption	72.2	79.4
Emissions/wastewater/waste	89.5	92.0
Biodiversity	35.3	28.7
Transport	84.2	67.0
Workplace/employment	84.2	92.2
Occupational safety and health	83.3	93.9
Training and development	100.0	91.9
Diversity and equal opportunity	89.5	75.8
Consumer protection	72.2	49.5
Child, forced and compulsory labour	77.8	45.0
Freedom of association/right to collective bargaining	84.2	53.9

Table 6.6 Measurement of sustainability issues' impact on company success and competitive advantage (n_{FRA} ranging from 17 to 19, n_{INT} ranging from 385 to 395)

	France	International average
Risks, with relevance for company success	68.4	47.1
Costs	66.7	54.0
Reputation/image/brand value	50.0	52.1
Revenue/sales/profits	55.6	44.0
Attractiveness as employer/job satisfaction	61.1	51.8
Innovation (products, processes etc.)	55.6	37.5
Efficiency/productivity	50.0	51.0
Business model innovation	35.3	26.5

Table 6.6 presents the measurement of sustainability issues' impact on company success and competitive advantage. Risks relevant for company success, costs, attractiveness as employer/job satisfaction, the impact in terms of revenue/sales/profits and in terms of business model innovation are more frequently measured in France compared to the international average. Conversely, reputation/image/brand value and efficiency/productivity are more frequently measured internationally than in France.

Discussion and implications of the implementation of sustainability by French companies

The implementation of sustainability in a corporate strategy is often evaluated in either the triple bottom line or the stakeholder expectation framework. The triple bottom line framework is based on three equally important axes of sustainable development: economic prosperity, environmental quality and social justice. The economic axis includes a firm's financial performance, its competitiveness and its ability to contribute to local economic development. The social axis covers the social impacts that a firm's activities have on all its stakeholders: employees (diversity, working conditions, remuneration, etc.), suppliers and clients (security, impact on product, etc.), the community and society in general (levels of nuisance, human rights, etc.). The environmental axis takes into account the impact of activities in terms of resource consumption, emissions of waste and pollutants and remedying environmental damages (Elkington 1999).

The second framework is based on the stakeholder concept developed by Freeman, who defined a stakeholder as an "individual or group of individuals with interests that may either affect, or be affected by, an organization" (1984:46). Starik (1995) argues for a broader definition which includes the natural environment as a crucial stakeholder.

Many of the guidelines of CSR management are based on stakeholder theory (Capron and Quairel-Lanoizelée 2004). According to this theory, if a company wants to integrate sustainability into its management, it must go beyond the expectations of its stockholders and take into consideration the needs of all its stakeholders. Different stakeholder typologies have been discussed in the literature. Carroll and Näsi (1997) distinguish between contractual stakeholders, who invest financial or human capital in the company, and disseminated stakeholders, who are impacted by the firm's activities without having a contractual relationship with it. The importance of a given social, environmental or economic issue is not the same for the firm and all its stakeholders. Companies must manage their relationships with their different stakeholders. For this purpose, they can adopt either a passive or a proactive approach. Whereas passive approaches (such as observation and informing stakeholders) are used by most companies, active approaches (such as stakeholder dialogue) are more frequently used in France than they are internationally. This result can be explained by the fact that the adoption of various regulations promoting the development of CSR in France was accompanied by a wave of new associations and informal working groups. In these groups are companies from different sectors who benchmark their practices and learning in sustainability management. More recently, these groups were opened to academics and researchers. This trend may explain the upsurge concerning the knowledge and the application of sustainability standards and guidelines in France. The relatively high awareness and application of ISO 26000 in France could also be explained by the active participation of the French delegation with a government representative at the debates within the International Organization for Standardisation on the ISO 26000 standard.

Apart from the necessity of managing stakeholder relationships, companies who want the sustainability strategies they implement to succeed must measure the impact that these strategies have on corporate success and competitive advantage. The ICSB results reported in this chapter indicate that French companies use measurement systems for some environmental issues (water and energy consumptions, materials, etc.) and some social issues (training, employment, etc.). These are generally items required by the NRE Act of 2001. In the coming years, as a result of the Grenelle 2 Law, there will be greater implementation of measurement systems for sustainability issues more connected to the company's core business.

In conclusion, French legislation has frequently intervened on questions of sustainability reporting and transparency. Accountability is inherent to the principle of responsibility. Both principles are deeply connected. A company cannot claim to be socially responsible if it does not report its CSR activities. According to Pérez (2003), good governance practices require that the representatives of an organisation disclose the consequences of their actions.

References

AccountAbility (1999) AA1000 framework. Standard, guidelines and professional qualification. The Institute of Social and Ethical AccountAbility, London

Bartlett JE, Kotrlik JW, Higgins CC (2001) Organizational research: determining appropriate sample size in survey research. Inf Technol Learn Perform J 19(1):43–50

Boasson C, Wilson A (2002) Développement durable et performance globale des entreprises: premières analyses de stratégies et pratiques. Rapport Entreprise et Personnel/CSR Europe, Décembre 2002

Bourion C (2006) Corporate social responsibility in proximity management practice. Revue Internationale de Psychosociologie 28:77–103

Capron M, Quairel-Lanoizelée F (2004) Mythes et réalités de l'entreprise responsable. La Découverte, Paris

Carroll AB, Näsi J (1997) Understanding stakeholder thinking: themes from a Finnish conference. Bus Ethics A Eur Rev 6(1):46–51

Di Maggio P-J, Powell WW (1983) The iron cage revisited: institutional isomorphism and collective rationality in organizational fields. Am Sociol Rev 48(2):147–160

Elkington J (1999) Cannibals with forks: the triple bottom line of the 21st century business. New Society Publishers, Gabriola Island

Freeman RE (1984) Strategic management: a stakeholder approach. Pitman, Boston

Gendron C (2000) Enjeux sociaux et représentations de l'entreprise. Revue du Mauss 15:320–325

Gray W-B (1987) The cost of regulation: OSHA, EPA and the productivity slowdown. Am Econ Rev 77(5):998–1006

Le Bas C (2004) Management responsable et nécessaire régulation globale. Économie et Humanisme 370:58–59

Orlitzki M, Schmidt F-L, Rynes S-L (2003) Corporate social and financial performance: a meta-analysis. Organ Stud 24(3):403–441

Pérez R (2003) On global responsibility in management. Corp Governance 3:78–89

Porter M, Kramer M (2006) The link between competitive advantage and corporate social responsibility. Harv Bus Rev 84(12):78–92

Robinson J-C (1995) The impact of environmental and occupational health regulation on productivity growth in U.S. manufacturing. Yale J Regul 12:346–387

Schaltegger S, Lüdeke-Freund F (2013) Business cases for sustainability. In: Idowu SO, Capaldi N, Zu L, Das Gupta A (eds) Encyclopedia of corporate social responsibility. Springer, Berlin/New York, pp 245–252

Schaltegger S, Harms D, Hörisch J, Windolph SE, Burritt R, Carter A, Truran S, Crutzen N, Ben Rhouma A, Csutora M, Tabi A, Kokubu K, Kitada H, Haider MB, Kim J-D, Lee K-H,

Moneva JM, Ortas E, Álvarez-Etxeberria I, Daub C-H, Schmidt J, Herzig C, Morelli J (2013) International corporate sustainability barometer: a comparative study of 11 countries. Centre for Sustainability Management, Lüneburg
Schaltegger S, Harms D, Windolph SE, Hörisch J (2014) Involving corporate functions: who contributes to sustainable development? Sustainability 6(5):3064–3085
Starik M (1995) Should trees have managerial standing? Towards stakeholder status for nonhuman nature. J Bus Ethics 14(3):207–217

Chapter 7
Corporate Sustainability Management in Large German Companies

Jacob Hörisch, Sarah Elena Windolph, and Stefan Schaltegger

Abstract This chapter introduces the results of the German sample of the International Corporate Sustainability Barometer (ICSB) and compares them with the results of the international average. German companies are generally perceived as comprehensively engaging in environmental and social issues. However, when analysing the results of the ICSB survey, it is striking how close German companies are to the international average rather than standing out. Nonetheless, some particularities can be found, for instance with regard to the awareness and application of employee-related sustainability management tools, which are more pronounced in German companies than on international average. There are also some negative deviations, such as the less close management of the freedom of association. Since German companies make up a very large share of the total sample (almost one-third), the results are not only compared to the overall international average but also to an alternative average excluding the German results. Possible explanations for the mediocre average positioning of the German companies are given and the potential for sustainability management is identified.

7.1 Introduction

7.1.1 The German Context

The German term for sustainability, *Nachhaltigkeit*, was first used in 1713 by the German forester and tax accountant Hans Carl von Carlowitz in a treatise on sustainable forestry (von Carlowitz 1713). This first recorded use of the term, some 300 years ago, is the basis of the claim that sustainability is originally a German

J. Hörisch (✉) • S.E. Windolph • S. Schaltegger
Centre for Sustainability Management (CSM), Leuphana University, Lüneburg, Germany
e-mail: hoerisch@uni.leuphana.de; windolph@uni.leuphana.de; schaltegger@uni.leuphana.de

S. Schaltegger et al. (eds.), *Corporate Sustainability in International Comparison*, Eco-Efficiency in Industry and Science 31, DOI 10.1007/978-3-319-06227-3_7,
© Springer International Publishing Switzerland 2014

concept (Wiersum 1995; Grober 2007). Keeping these historical roots in mind, it does not surprise that sustainability is an important issue in societal, political and economic debates in Germany today.

Germany is a highly industrialised, economically developed and export-oriented economy. It has a high gross domestic product (GDP) (in absolute terms as well as per capita), a very large and increasing share of which (41.5 %) is generated by exports (The Economist 2008). With a relatively high rate of income redistribution, it fulfils the main criteria of a typical continental European welfare state (Schmidt 2008). Its political system is characterised by a large number of federalist mechanisms for sharing power between the federal, state and local levels. Environmental and sustainability issues, like greenhouse-gas emissions, are mostly regulated by the federal government. In 1999, the federal government introduced a national sustainability strategy, including numerous corporate sustainability (CS) objectives. Progress in implementing this national sustainability strategy is published in regular reports (Federal Statistical Office of Germany 2012). In the context of corporate social responsibility (CSR) and CS, the German Federal Ministry for the Environment, Nature Conservation and Nuclear Safety is also of core importance, since it prepares the legal framework (Leal Filho and Pawlak 2009). It is also responsible for coordinating the phase-out of nuclear power, which the German federal government aims to achieve by 2022. Besides national regulations, the European Union (EU) has an important influence on the framework conditions for corporate sustainability management in Germany. One of its most important initiatives was doubtlessly the introduction of the Eco-Management and Audit Scheme (EMAS) in 2001.

In addition to the legal framework, corporate networks play a crucial role for sustainability management in Germany. They facilitate the exchange of views and knowledge on current developments, support best practices through awards and awareness campaigns and help establish common standards, for example by collaborating with the government. Among the most important networks are Econsense, which was founded in 2000 on an initiative of the Federation of German Industries (BDI), and CSR Germany, the national network of CSR Europe, which is coordinated by the BDA (Confederation of German Employers' Associations). Last but not least, BAUM e.V. (German Association of Environmental Management), founded in 1984, is the oldest German business association in the field of sustainability, originally focusing on environmental management.

Besides the companies themselves and government regulations, consumers obviously have a key role to play in supporting new, more sustainable business practices. The World Values Survey (2009) documents that in the German public the awareness of many environmental issues is above the international average. However, when it comes to the willingness to pay for environmentally-friendly products or to possible tax increases for additional government expenditures to protect the environment, Germans tend to be more reluctant than the citizens and consumers in other countries. Nevertheless, Armingeon et al. (2010) found the share of votes for green parties to be among the highest in Germany compared to all other OECD member states, supporting the observation that general environmental awareness is relatively high in Germany.

The business environment described above with regard to legislation, corporate networks and consumer attitudes indicates that, on the one hand, there is potential for leadership in corporate sustainability. On the other, some factors like the relatively low willingness to pay for green products and services pose important challenges for corporate sustainability in Germany. In their analysis of 19 country-specific approaches to CSR, Leal Filho and Pawlak (2009) find that many German companies accept the challenge of corporate sustainability. In comparison to the other countries in the study, the authors emphasise the relatively high importance for German companies to link CSR activities to their core business. However, at the same time, based on their investigation of CSR examples in large German companies, the authors find philanthropic activities to be of high relevance in Germany. Similar tendencies are identified in other large-scale surveys on sustainability management in Germany. The Corporate Sustainability Barometer (CSB) found in 2010 as well as in 2012 that the vast majority of large German companies claim to link sustainability to their core business (84.8 and 94.1 %, respectively). While in 2010 some companies mistakenly cited philanthropic activities as evidence of such a connection, this share significantly decreased in 2012 (Schaltegger et al. 2010; Schaltegger et al. 2012).

7.1.2 The German Sample

The German survey is part of the International Corporate Sustainability Barometer coordinated by the Centre for Sustainability Management in Lüneburg. To analyse and internationally compare German corporate practices of sustainability management in more detail, the largest 500 German companies as well as the 50 largest banks and 30 largest insurance companies (excluding subsidiaries) formed the basic population of the German Corporate Sustainability Barometer in 2012. Of this basic population, 383 companies received the questionnaire, 152 of which participated in the survey (response rate of 39.7 %). Comparing the key characteristics of the German sample with the international dataset reveals some important differences (see also Schaltegger et al. 2013).

First, German companies tend to be larger (on a revenue basis) than their international peers. While only 19.7 % of the German companies are characterised by annual revenues of less than €1,500 million, this is true for 31.0 % of the companies in the international sample. In addition, 20.4 % of the companies in the German sample have annual revenue higher than €50,000 million, compared to 9.8 % in the international sample. Similar but less distinct differences can be found concerning the average number of employees.

The organisational structure also differs between the German sample and the international average. Whereas in Germany 28.7 % of the responding companies are family-run, this holds true for no more than 19.3 % in the international group. Further differences include the share of non-domestic sales. German companies tend to be more export-oriented than their international peers. Whereas nearly 60 % of the

German companies earn more than 40 % of their sales abroad, this is true for only 48.5 % of the companies in the international sample. This is somewhat surprising, since the ICSB survey reveals that high shares of non-domestic sales are usually found in smaller countries. With regard to the companies' core business activities, no significant deviance can be found for the German sample.

Lastly, it is worth emphasising that the German companies make up 32.5 % of the international dataset (152 of the total 468 surveyed companies), meaning that they exert a disproportional influence on the international average. This is why, for some of the questions analysed in the following section, the German figures are compared with an alternative international sample – the data of all investigated countries *excluding* Germany. This allows meaningful deviations in the German sample to be identified and existing differences to be highlighted.

7.2 Analysis: The German Results

7.2.1 Intention

Stakeholders play a crucial role in motivating companies to engage in sustainability management (Bansal and Roth 2000; Black and Härtel 2004; Epstein 2008; Lacy et al. 2010). Similar patterns can be found for the international and the German samples, since in both cases NGOs/environmental/social organisations, the media/public and international authorities are perceived to be the most promoting stakeholders, whereas the influences of trade unions, insurance companies and banks are more neutral (see Fig. 7.1). Minor differences can be found for the influence of media and the public, which are perceived as even more promoting in Germany, and the local community, which is rated as less important.

Besides stakeholders, several *other factors,* such as corporate philosophy, legislation or know-how within the company, influence the likelihood an organisation will implement corporate sustainability. Again, similar patterns can be observed in Germany and on international average. In both samples the commitment by top management is the most important promoting factor, whereas a lack of personnel capabilities inhibits corporate sustainability most. Although the ranking of issues is similar, differences in the degree of promoting or inhibiting can be observed. The self-commitment by business, for example, is perceived to have a far less promoting effect in Germany, and more companies evaluate it as an inhibiting factor than on international average. In contrast, a lack of personnel capacities and a lack of government incentives are less frequently assessed as important inhibiting factors in Germany.

Companies can also be motivated to pursue corporate sustainability because they *expect positive impacts* on corporate success. Overall, German companies expect similar effects as their international peers. However, in Germany business model innovation is less frequently stated to be an objective of corporate sustainability.

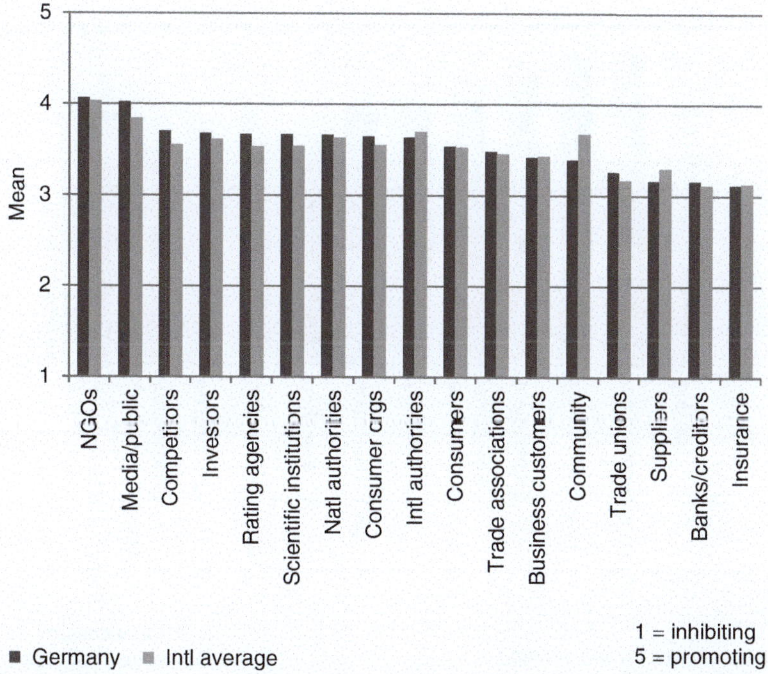

Fig. 7.1 Influence of external stakeholders on corporate sustainability in Germany (n_{GER} ranging from 128 to 146, n_{INT} ranging from 393 to 450)

The different motivations for a company's sustainability engagement may also serve to explain which *issues* a company manages. Comparing the German sample with the international sample reveals that nearly all sustainability issues are managed slightly less closely in Germany (see Fig. 7.2). The difference is largest for freedom of association/right to collective bargaining and water consumption. The latter might be explained by the fact that water shortages are very rare in Germany. Therefore, it is also unsurprising that stakeholders only rarely demand companies to substantially reduce water consumption.

For both the international and the German samples, a correlation can be observed between *stakeholder requirements* and *issues managed*. Thus, the relatively low engagement of German companies for freedom of association/right to collective bargaining may at least partly be explained by stakeholder requirements regarding this issue far below the international average. Similarly, lower values for stakeholder requirements in Germany can be observed for consumer protection and occupational health and safety.

Investigating the degree of *stakeholder criticism* on these aspects and analysing its development over the 2 years preceding the survey reveals some surprising findings. For all aspects the share of companies perceiving a decrease in stakeholder criticism of a particular issue is lower in Germany than in the international sample.

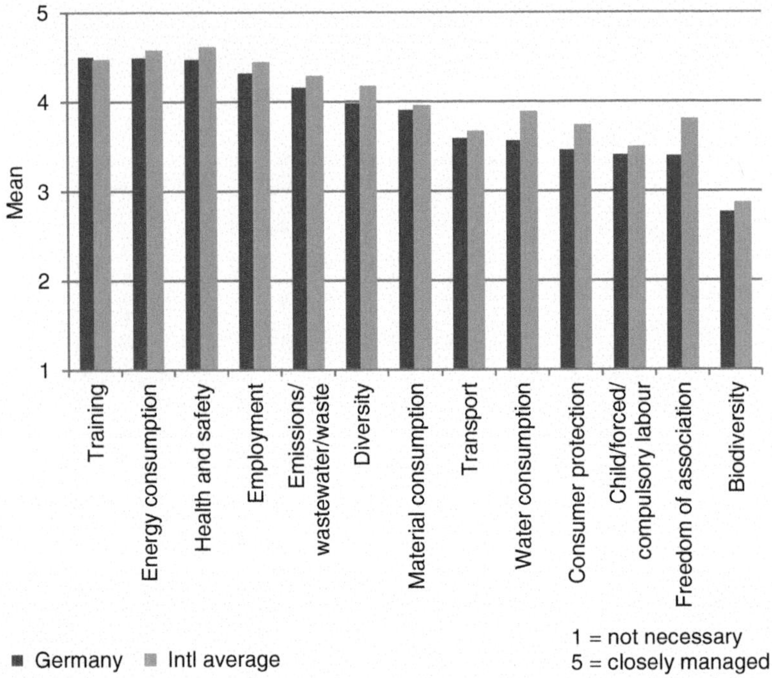

Fig. 7.2 Managed sustainability issues in Germany (n_{GER} ranging from 140 to 150, n_{INT} ranging from 442 to 463)

Additionally, for all issues a larger share of companies recognises an increase in stakeholder criticism in Germany than in the international sample. These differences are largest for diversity/equal opportunity, energy consumption, consumption of materials, and emissions/waste/wastewater. Different explanations are possible for this phenomenon. First, stakeholders could have become increasingly aware of these issues over the 2 years preceding the survey. Given the discussions in Germany on introducing a female quota in boardrooms this seems plausible for the case of diversity and equal opportunity. Similarly, it could be argued that the German debate on phasing-out nuclear power and the energy transition towards renewables has increased stakeholder awareness of corporate energy consumption. Alternatively, insufficient company performance could have led to increased criticism. To identify the specific reasons, more detailed research on a company-specific level is needed.

7.2.2 Integration

The German results show that German companies tend to integrate sustainability into their *core business* a little more frequently than on international average.

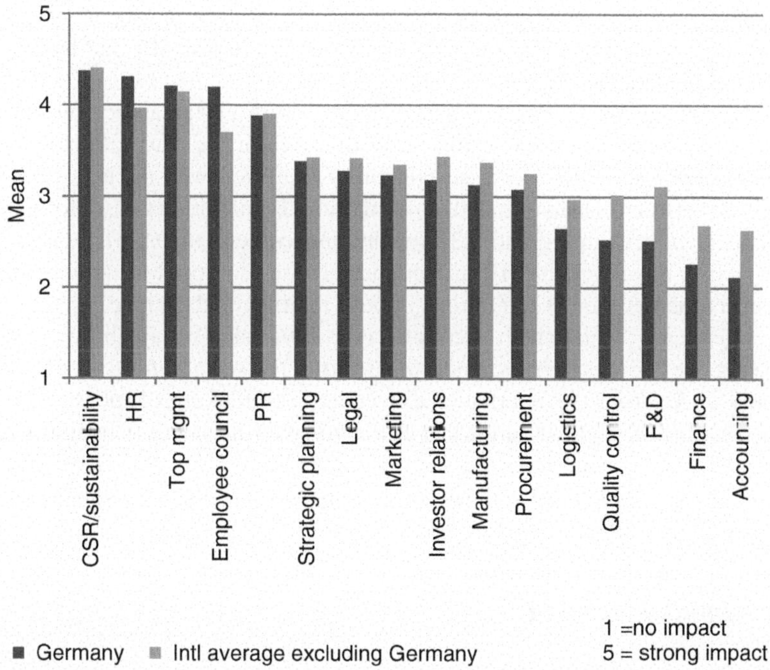

Fig. 7.3 Organisational units impacted by social issues in Germany (n_{GER} ranging from 107 to 149, $n_{INT\ excl.\ GER}$ ranging from 214 to 308)

In contrast to companies in some other countries, none of the German companies state that they strictly separate sustainability from their core business. In addition, the share of companies stating they connect sustainability to some, most or all parts of their core business is slightly higher in the German sample (96.0 %) than on international average (92.8 %). The most significant difference can be found for companies connecting sustainability to "most segments of their core business" (45.0 % in Germany versus 40.7 % on international average). However, fewer companies in Germany indicate they consistently connect sustainability to "all segments" of their core business than on international average. Overall, the differences can be considered relatively small.

Whereas the surveyed German companies evaluate the extent that *organisational units* are affected by *environmental issues* only slightly differently from the international average, more significant differences can be found for the impact of *social issues*. In Germany almost all organisational units are assessed as less influenced by social issues than on international average. These differences are particularly large (about 0.3 or higher) for research & development, financial & management accounting, quality control and finance. Calculating the alternative international average *excluding* the German companies, the difference for these units is higher, between −0.42 and −0.6 (see Fig. 7.3). The extent to which the employee council is affected by social issues is, however, higher in Germany (+0.5).

Regarding environmental issues, particularly financial & management accounting, the legal department and the employee council are less affected in Germany than on international average.

Overall, the *engagement of functional units* with sustainability measures in German companies does not significantly differ from the international average. Some units engage a little less in sustainability measures, with the largest negative differences between Germany and the overall sample amounting to 0.24 for financial & management accounting and 0.22 for the finance department. The engagement of research & development, HR and public relations/corporate communication in German companies, however, is slightly above international average.

The frequency with which German companies take sustainability measures addressing the various *drivers of business cases for sustainability* does not significantly differ from the international average. The ranking of measures is also almost the same. A slight variation can be identified for measures serving to manage environmental and social risks (e.g. health care at the workplace). These measures are taken marginally less often in German companies than on international average.

7.2.3 Implementation

A comparison of the *management of stakeholder relations* in Germany and on international average does not provide a clear overall picture. The share of German companies using different stakeholder relations at least on a *case-specific basis* is larger for some of the stakeholder relations (observing, consideration in decision-making process, cooperating) but smaller for others (such as delegating decision-making authority). The share of German companies managing the different stakeholder relations *in most cases or in general,* and not just on a case-specific basis, is smaller than on international average. The difference is largest for 'dialogue with stakeholders' (−7.3 %).

The *awareness and the application of sustainability management tools* are slightly different in German companies than on international average. First, in German companies several tools are *applied* above average (with differences above 10 %), namely proposal system (applied in 84.1 % of the German companies, which is +28.2 % above the international average), human resource control (+18.8 %), eco-indicators (+18.2 %), social/cultural sponsoring (+17.8), further education (+16.1 %), eco-audit (+16.1 %), eco-circle (+15.3 %), social indicators (+15.1 %) and quality circle (+14.7 %; see Fig. 7.4). All of these tools are also *known* in a larger share of the German companies than on international average.

Additionally, further sustainability management tools are better *known* in German companies than on international average (differences above 10 %), i.e. eco-control (+14.3 %) and sustainability control (+12.2 %), and they are both also applied slightly more frequently.

However, for some tools their *application* is *lower* in German companies than on international average (differences above 10 %): corporate giving (−19.5 %),

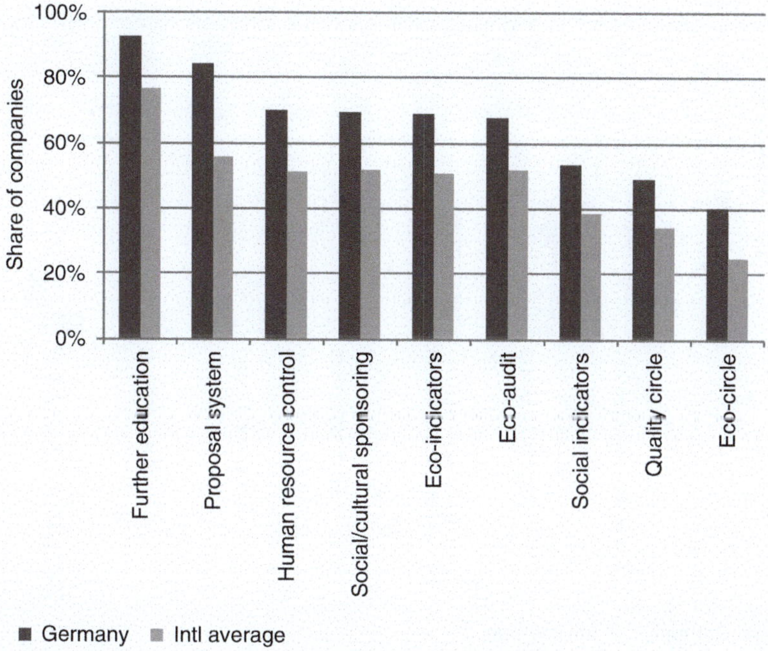

Fig. 7.4 Sustainability management tools applied more in Germany than internationally (per cent; $n_{GER} = 151$)

green purchasing (−19.1 %), community advisory panel (−15.8 %), environmental statement (−11.8 %), environmental information system, environmental accounting (both −11.3 %), corporate/employee volunteering (−10.5 %; see Fig. 7.5). All these tools are also less known in the German companies than on international average.

In sum, tools addressing *employee* issues seem to be particularly popular in German companies, e.g. proposal system, human resource control and further education. Moreover, *indicators* and tools serving to *measure and steer* sustainability issues are more widespread in Germany.

For a selection of tools the survey examined how long companies have known them (in case a company stated it knows the tools). The respondents could choose between the categories '2 years or less' and 'more than 2 years'. With the exception of social benchmarking and green supply chain management, the share of the listed tools known 'more than 2 years' by German companies was higher than on international average.[1] This indicates that the awareness of sustainability management tools is institutionalised to a larger extent in Germany than among other countries.

[1]This question was not surveyed in the UK and Hungary. The international average here only consists of the information from the companies of nine countries.

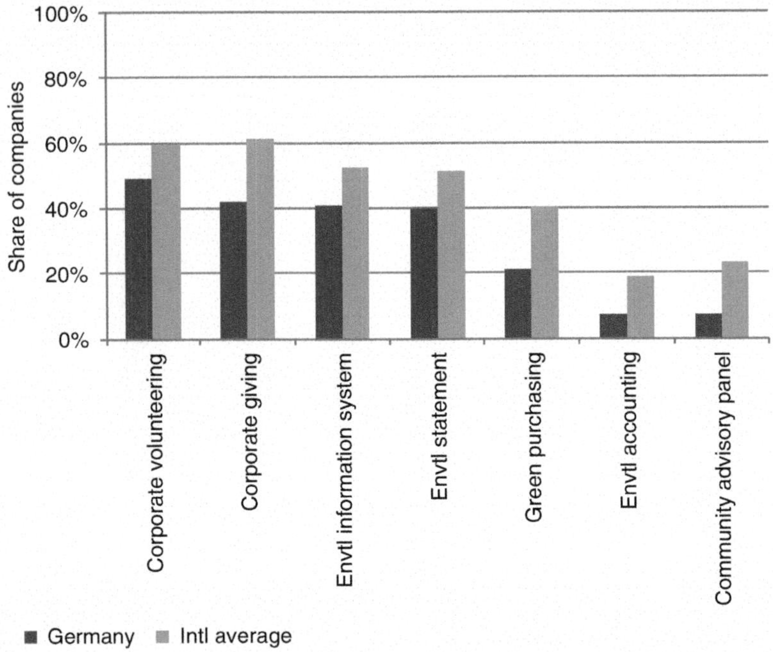

Fig. 7.5 Sustainability management tools applied less in Germany than internationally (per cent; $n_{GER} = 151$)

In the surveyed German companies the *awareness of sustainability standards or guidelines* is slightly above international average, since 8 of the 12 standards are known more frequently in the German sample. Yet, the *application* of standards or guidelines in German companies is below average for 9 of the 12 standards. However, the differences are relatively small with regard to both the awareness (maximum difference is +6.0 % for EFQM) as well as the application (maximum difference is −10.1 % for ISO 26000).

7.3 Discussion and Conclusions

Overall, the comparison with the international average shows that for some issues the surveyed German companies yield higher or more positive results, whereas for other issues the German results are found to be below average. The strongest *positive* deviations can be found for the awareness and application of several sustainability management tools, especially employee-related tools as well as indicators and tools serving to measure and steer sustainability issues. Furthermore, the finding by Leal Filho and Pawlak (2009) on the integration of sustainability issues (see Sect. 7.1) is supported, since German companies do tend to connect sustainability to their core business somewhat more systematically than companies in most other countries investigated. Leal Filho and Pawlak's observation on the importance of corporate philanthropic activities, however, cannot be confirmed, as the ICSB data does not show a large importance of corporate philanthropy in Germany. Comparing the German CSB data from 2012 to 2010 even reveals a decrease in companies stating philanthropy as an example for connecting sustainability with their core business (see Sect. 7.1).

Negative deviations can be identified, e.g., for the issue freedom of association, which is less closely managed in German companies than on international average. However, this result might also indicate that in Germany there is simply not as much need for action regarding this issue, since trade unions are traditionally well-established in a number of economic sectors. Notably, German companies less frequently perceive a decrease and more frequently perceive an increase in stakeholder criticism than on the international average. This may indicate that German companies achieve less progress than the other countries for social and environmental issues – and/or that stakeholders are more critical in this country.

Yet, for most of the aspects discussed, the differences are relatively small and the German companies are generally close to average. This is somewhat surprising, since several aspects described in the introduction gave reason to expect above-average corporate sustainability engagement. Among these factors are the high share of very large companies (see Brammer and Pavelin 2006; Gallo and Christensen 2011), the importance of green parties as well as the longstanding tradition of sustainability and environmental concern in Germany. However, a closer look at the German context reveals additional circumstantial factors which may explain the mediocrity of the German results. First, the strong export orientation (cf. Federal statistical office of Germany 2013) may explain why many German companies are more influenced by international than by national developments. Second, it has to be kept in mind that German companies make up almost one-third of the international sample of all countries. Therefore, they heavily influence the international average. Comparing the German survey results with the adjusted international average (excluding Germany), however, leads to the same result that companies in Germany represent the international average for most corporate sustainability issues.

To sum up, compared to the other countries analysed, German companies do not stand out particularly positively. However, German companies also do not score particularly low for any of the aspects analysed, whereas a lot of the other countries show highlights in some areas but at the same time deficiencies in others. In summary, the majority of German companies does quite well with regard to sustainability as no significant shortcomings can be identified.

One possible way to overcome the German mediocrity with regard to corporate sustainability would be to reduce reliance on government regulation (see World Values Survey 2009). Both companies and consumers would then need to become actively involved in embracing the business opportunities of sustainability management.

References

Armingeon K, Careja R, Engler S, Gerber M, Leimgruber P, Potolidis P (2010) Comparative political data set III. http://www.ipw.unibe.ch/content/team/klaus_armingeon/comparative_political_data_sets/index_ger.html. Accessed 07 Mar 2013

Bansal P, Roth K (2000) Why companies go green: a model of ecological responsiveness. Acad Manage Rev 43(4):717–736

Black LD, Härtel CEJ (2004) The five capabilities of socially responsible companies. J Public Aff 4(2):125–144

Brammer S, Pavelin S (2006) Corporate reputation and social performance: the importance of fit. J Manag Stud 43(3):435–455

Epstein MJ (2008) Making sustainability work. Best practices in managing and measuring corporate social, environmental and economic impacts. Greenleaf/Berrett-Koehler, Sheffield/San Francisco

Federal Statistical Office of Germany (2012) Sustainable development in Germany. Indicator report 2012. Federal Statistical Office, Wiesbaden

Federal Statistical Office of Germany (2013) Foreign trade. Export propensity, Germany. https://www.destatis.de/EN/FactsFigures/NationalEconomyEnvironment/ForeignTrade/ForeignTradeIndicators/Tables/ExportPropensity.html. Accessed 07 Mar 2013

Gallo PJ, Christensen LJ (2011) Firm size matters: an empirical investigation of organizational size and ownership on sustainability-related behaviors. Bus Soc 50(2):315–349

Grober U (2007) Deep roots – a conceptual history of 'sustainable development'. Wissenschaftszentrum Berlin für Sozialforschung (WZB), Berlin

Lacy P, Cooper T, Hayward R, Neuberger L (2010) A new era of sustainability. CEO reflections on progress to date, challenges ahead and the impact of the journey toward a sustainable economy. http://www.accenture.com/SiteCollectionDocuments/PDF/Accenture_A_New_Era_of_Sustainability_CEO_Study.pdf. Accessed 8 May 2014

Leal Filho WL, Pawlak PS (2009) Federal republic of Germany. In: Idowu SO, Leal Filho W (eds) Global practices of corporate social responsibility. Springer, Heidelberg, pp 61–71

Schaltegger S, Hörisch J, Windolph SE, Harms D (2012) Corporate sustainability barometer 2012. Centre for Sustainability Management, Lüneburg

Schaltegger S, Windolph SE, Harms D (2010) Corporate sustainability barometer 2010. Centre for Sustainability Management, Lüneburg

Schaltegger S, Harms D, Hörisch J, Windolph SE, Burritt R, Carter A, Truran S, Crutzen N, Ben Rhouma A, Csutora M, Tabi A, Kokubu K, Kitada H, Haider MB, Kim J-D, Lee K-H, Moneva JM, Ortas E, Álvarez-Etxeberria I, Daub C-H, Schmidt J, Herzig C, Morelli J (2013) International corporate sustainability barometer: a comparative study of 11 countries. Centre for Sustainability Management, Lüneburg

Schmidt MG (2008) Germany: the grand coalition state. In: Colomer JM (ed) Comparative European politics, 3rd edn. Routledge, London/New York

The Economist (2008) Pocket World in Figures. 2009 edition. Profile Books Ltd/The Economist, London

von Carlowitz HC (1713) Sylvicultura Oeconomica. Haußwirthliche Nachricht und Naturmäßige Anweisung zur Wilden Baum-Zucht, 2nd edn. Braun, Leipzig

Wiersum KF (1995) 200 years of sustainability in forestry: lessons from history. Environ Manage 19(3):321–329

World Values Survey Association (2009) World values survey 1981-2008 official aggregate v.20090901. www.worldvaluessurvey.org. Accessed 15 Feb 2013

Chapter 8
Sustainability Management in Hungary

Maria Csutora, Sandor Kerekes, and Andrea Tabi

Abstract In Hungary the culture for sustainability management is quite diverse. Even large companies are usually subsidiaries or suppliers to large multinational companies and the influence of buyers and owners is substantial. As a result there are a number of different traditions in sustainability management in the country and correspondingly relatively few typical cultural attributes. The impacts of investors and shareholders are given high scores in the survey, while consumers are given a low score in motivating companies to pursue sustainability management. Community also has little influence because community involvement in sustainability management does not have a long tradition in Hungary. Thus two-way communication and participative methods of stakeholder management are less common. The responses show that Hungarian companies manage most environmental issues, especially emissions, more closely than the international average. Sustainability management tools are broadly known and applied in the country, and the general satisfaction with the number and level of tools indicates that there is no need for further development here. Basically, Hungarian companies are more sceptical towards the benefits of the implementation of corporate sustainability. This is especially true with regard to the prospective positive impacts on cost reduction, innovation, employee motivation as well as enhancing and safeguarding corporate reputation. In sum, Hungarian companies have already demonstrated expertise in most fields of sustainability management, but the development of a more participative collaboration with both internal and external stakeholders in sustainability management is still needed.

M. Csutora (✉) • A. Tabi
Sustainability Indicators Research Centre, Corvinus University of Budapest, Budapest, Hungary
e-mail: maria.csutora@uni-corvinus.hu; andrea.tabi@uni-corvinus.hu

S. Kerekes
Department for Environmental Economics, Corvinus University of Budapest, Budapest, Hungary
e-mail: sandor.kerekes@uni-corvinus.hu

S. Schaltegger et al. (eds.), *Corporate Sustainability in International Comparison*,
Eco-Efficiency in Industry and Science 31, DOI 10.1007/978-3-319-06227-3_8,
© Springer International Publishing Switzerland 2014

8.1 Introduction

8.1.1 The Hungarian Context

During the last 15 years the share of large companies in the Hungarian economy has halved. This huge structural shift has created a significant level of unemployment in the country and the rapid growth in the number of SMEs has not been able to absorb the 600,000 newly unemployed workers (Boda et al. 2010). When evaluating sustainability issues in Hungary in particular then, we cannot ignore this fact and we must take into consideration a possible trade-off between sustainability, competitiveness and employment possibilities. The main question is whether a company is able to harmonise these goals. Societal actors are more patient with achieving sustainability goals and less critical of deficits in environmental protection if the company is a major provider of employment, pays above average wages and reinvests profits within the country (Kerekes et al. 2006). The vast majority of large Hungarian corporations are affiliated to large multinationals while investments are typically greenfield investments. Many of them are business groups or big supermarkets, but the majority are in the manufacturing sector, which is why Hungary has become one of the main manufacturers in Europe. This rapid but distorted development has become apparent in the fact that the majority of large manufacturers are in the automobile sector. Whereas the car manufacturing sector did not exist in Hungary before 1990, in 2012 one-fifth of industrial production and one-fifth of Hungarian exports originated in this sector (Gelei 2003; Klauber 2008).

The car manufacturing sector has a very positive influence on eco-efficiency but this is not always the case. It is especially doubtful at the beginning of the supply chain when transportation costs have only a marginal influence on procurement processes although their global environmental impact is quite large. For example, an aluminium manufacturer processing tens of thousands of tons may decide not to purchase even small amounts from an aluminium smelter located nearby, because the purchase of aluminium cast on the world market as a big consumer offers more advantages than purchasing it from a company geographically close to it – even when this heavy cast must be transported several thousand kilometres and means a high environmental burden. The sustainability reports of these multinationals always include information about recycling but almost never about the transportation distances.

In Hungary the culture for sustainability management is quite diverse due to its openness to the world market. Even the largest companies, apart from the few Hungarian multinationals, are usually subsidiaries of or suppliers to large multinational companies, or sometimes both. Their culture for sustainability does reflect the culture of the shareholders and the expectations of business partners. For example, according to the official list of EMAS-registered sites held by the Ministry of Rural Development, all foreign-owned companies with EMAS certification have a German origin (Ministry of Rural Development 2013). Japanese companies are more interested in ISO 14001 and all members of the Hungarian-Japanese Economic

Club hold ISO 14001 certification.[1] Service companies and companies labelled "dirty" by the public, e.g. in the chemical sector, are corporate social responsibility (CSR) oriented and tend to implement a standardised environmental management system. Companies that are suppliers to the automotive industry might prefer instead integrated quality control and environmental management systems (Kerekes et al. 2003; Csutora and Harangozo 2009; Harangozo et al. 2010). In most companies, with the exception of those with a CSR focus, the environmental manager holds an environmental engineering degree. The strong engineering and technological orientation of environmental managers is perhaps an attribute shared by most companies in the country. Thus, different traditions for sustainability management exist side by side in the country and there are few typical Hungarian cultural attributes.

8.1.2 The Hungarian Sample

The top Hungarian 200 companies were approached, as only these companies meet the survey criteria. Of the 200 companies, 85 agreed to be involved in the survey, with 28 of them actually responding to the survey in the end. The survey is part of the International Corporate Sustainability Barometer coordinated by the Centre for Sustainability Management in Lüneburg (see Schaltegger et al. 2013).

It is interesting that although only the largest companies were surveyed the average number of employees in the Hungarian sample is still the lowest compared to the other countries surveyed. While 46.2 % in the international sample have 10,000 or more employees, in Hungary only 7.4 % belong to these categories, all the other companies being smaller. Both the size of the economy and the average size of the companies are below the international average in terms of annual revenue and number of employees. Almost 60 % of the companies surveyed in Hungary had 1,001–10,000 employees in the last financial year. One-third of the companies had fewer than 1,000 employees and the rest over 10,000. The majority (64.3 %) of companies had revenue in the last financial year in the range from €50 to €500 million and the rest of them account for more than €500 million per year. In Hungary an above-average percentage (55.6 %) of companies compared to the international sample have a share of non-domestic sales of greater than 80 %.

More than half (55 %) of the Hungarian economic organisations constitute joint ventures and the rest (45 %) consist of entrepreneurships, with 36 % of existing enterprises being companies with legal entity and only 1 % shareholder corporations (KSH Database 2013).

Companies differ in their core businesses, with almost 40 % of the companies surveyed belonging to the group 'industry, capital goods & building', 25 % to 'consumer goods, trade & logistics' and around 35 % of the companies being almost

[1]According to the information released on the homepages of the companies listed by the Hungarian-Japanese Economic Club.

evenly split between 'finance & services' and 'commodities, auxiliary material, energy, chemical & pharmaceutical'. Significant differences from the international sample can be found with regard to core business, e.g. that the industry, capital goods & building sector is very large in Hungary (39.3 % versus 22.6 % in the international sample), whereas the finance & services sector is relatively small in the Hungarian sample (17.9 % versus 32.1 % on international average).

8.2 Analysis: Comparison to the International Average

8.2.1 Intention

The impact of internal organisational units on the implementation of sustainability among Hungarian companies is above the international average; the values given for different internal stakeholders are higher or similar. The only exception is purchasing/procurement, where these impacts lag behind the international sample. Green procurement is not widespread within the country and especially not in the B2B companies that dominate our sample (Magerholm Fet and Zilahy 2011).

With regard to external stakeholders, the impact of suppliers is given a surprisingly high evaluation. A possible explanation might be the high share of companies in the sample from the automotive industry, which closely monitors environmental impacts in the supply chain. Another analysis of 467 Hungarian companies showed that outstanding environmental management performance in the automotive sector is most often supply-chain driven while other sectors tend to be characterised by company culture or pressure-driven strategies (Csutora et al. 2006). Before 2012 the Hungarian tax system encouraged innovation-related collaboration between companies and academic institutions by means of a so-called innovation contribution or 'R&D tax',[2] which had a positive impact on the sustainability practices of companies. Market-oriented stakeholders such as consumers, consumer organisations and rating agencies score significantly lower in Hungary than on international average.

8.2.1.1 Motivation

Compared to the international sample, consumers/end users are given low scores in motivating sustainability management in Hungary. Most Hungarian companies are in B2B industries and are thus not directly connected to end users. Community does

[2]Act XC of 2003, approved by the Hungarian Parliament on November 10, 2003, established the Research and Technological Innovation Fund, which provides stable and reliable financing for RTDI activities (http://www.nih.gov.hu/english/hungarian-innovation-act/overview-of-the-research).

not have much influence either, because community involvement and participative methods in sustainability management do not have a long tradition in the country. The paternalistic legacy of the previous socio-political regime still has an impact on current attitudes towards stakeholder involvement (Bodorkós and Pataki 2009).

National and international authorities/legislators are also evaluated significantly lower than in other countries. Scientific institutions are given a higher score than on international average. Up to 2012, the ‘innovation contribution’ could be waived if the company hired R&D services of the same value from scientific institutions. This regulation encouraged closer collaboration between companies and scientific institutions and sustainability-related innovation service was often part of this framework. Scientific institutions thus had a relatively high influence on environmental R&D in companies. From 2012 on however the innovation contribution must be paid as a tax without any possibility of exemption or allowance.

Basically, Hungarian companies are more sceptical towards the benefits from the implementation of corporate sustainability. This is especially true for prospective positive impacts of cost reduction, innovation, employee motivation as well as enhancing and safeguarding corporate reputation. Scepticism can be explained by the background of leading sustainability managers, who are mainly engineers and emphasise technology and put less emphasis on social and economic issues.

It is interesting that in Hungarian companies sustainability issues are less expected to have a positive effect on motivating employees, while in the international sample the expected positive impact of sustainability management on employee commitment is evaluated as high. This finding is probably due to the fact that the approach towards environmental and social issues is less participatory in Hungary, which is confirmed by other questions in this survey, too.

Companies list legislation and the lack of government incentives, e.g. tax advantages, as major hindering factors for the implementation of sustainability management. Especially the lack of government incentives is evaluated as an important inhibiting effect on improving corporate sustainability, being almost one point higher than the international average on a five-point scale. This might be due to the changes in the tax system described above. The surveyed companies have a high opinion of their capacity to manage the complex issue of sustainability and of their know-how in general. Lack of know-how in the company and the methodological complexity of sustainability management are the least *inhibiting* factors of the inhibiting factors queried, and they are evaluated as *less* inhibiting than on international average. The difference is around 0.3 in both cases. This might be a culture-specific value, as Hungarian engineers tend to rate their skills and capacity highly.

8.2.1.2 Issues

According to the responses, Hungarian companies manage environmental issues more closely than in the international average, especially energy, water and material consumption as well as emissions, waste and wastewater. The median is 5 for all issues, and the mean is above the international average. Among the different impacts of sustainability management on company success, activities associated with cost savings, e.g. energy savings, are mostly closely monitored in Hungary. Sometimes, however, savings are accounted for and recorded by manufacturing or maintenance departments rather than by sustainability units. For example, innovation leading to energy or water savings might be initiated by departments as diverse as production, maintenance or sustainability (Széchy 2012). Biodiversity is given the least attention. Social issues are managed somewhat less than on international average, especially consumer protection as well as diversity and equal opportunities. Although diversity issues are less closely managed than in the international sample, a previous survey shows that gender balance in management positions has already reached the European average (Nagy 2005). The difference is not high, though, compared to the international sample (−0.35).

Freedom of association/right to collective bargaining seems to be paid more attention among Hungarian companies, which might be due to a cultural legacy from the socialist period.

While in the international sample the extent of stakeholder criticism has increased during the 2 years preceding the survey, no such tendency is observed among Hungarian companies. Stakeholder criticism has remained mostly unchanged or no criticism is observed. How much the lack of criticism reflects the opinions of stakeholders – or if it is mainly due to the lack of bilateral communication – would need further research. Hungarian companies watch stakeholder opinion less closely and rely more on their perception of stakeholder opinions. Based on the survey responses, a gap seems to exist between the perceived expectations of external stakeholders and the engagement of companies in active dialogue with their stakeholders when managing sustainability issues. When asked how they manage stakeholder relations, Hungarian companies score lower than their international peers on participative methods, including observing stakeholders, dialogue with stakeholders, involvement and consideration in decision-making processes or networking with stakeholders to develop joint solutions. The percentage of Hungarian companies that do not use these participatory methods is higher than in the international sample. The communication between Hungarian companies and their stakeholders is thus less systematic and tends to feature a case-by-case approach.

8.2.2 *Integration*

8.2.2.1 Connection to Core Business

Only 20.0 % of the companies surveyed consistently connect all segments of their core business to sustainability, which is lower than the international average (27.8 %). Several Hungarian companies connect sustainability to some of the segments of their core business (24.0 %) and the largest share claims that sustainabilityis integrated into most segments (44.0 %) of their companies. Separation of sustainability issues and sustainability management from business issues and business management can be observed throughout the surveyed companies. In our opinion this might make communication between financial or accounting departments and sustainability managers difficult (see also Schaltegger et al. 2012).

8.2.2.2 Involvement of Organisational Units

Similar to the international results, top management, public relations/corporate communication (PR) and CSR seem to be the most important organisational units for the implementation of corporate sustainability. Less significant units are finance, financial & management accounting and the employee council.

R&D, manufacturing, top management and the CSR department are most strongly impacted by environmental issues. These departments are usually directly involved in sustainability management innovation. Marketing, procurement, PR, logistics, quality control and strategic management also score high, and other organisational units are also impacted above the international average. The same applies to societal issues, which also impact some organisational units in Hungarian companies more strongly than on international average.

An interesting discrepancy can be observed for the manufacturing unit. Although this organisational unit is among those impacted the most by environmental issues according to the respondents, its engagement for sustainability measures is low compared to the international sample and to other departments. For instance, marketing, CSR and human resources (HR) units are more engaged in sustainability measures according to our survey. These results can be interpreted such that when manufacturing aspects and sustainability aspects conflict with each other, manufacturing departments tend to prioritise the former. Manufacturing departments do care about sustainability to a certain extent, but their engagement is at a lower level when compared to marketing, CSR or HR departments.

8.2.2.3 Drivers for Sustainability

Sustainability measures are implemented to the same extent as in the international sample in the fields of developing new customer and business segments, promoting employee motivation and more efficient use of resources. External communication of environmental and social activities (e.g., sustainability reporting) and environmental and social-oriented cost management (e.g., using cost-effective recycling products) lag behind other countries. This finding is consistent with responses to other questions that indicate a low level of stakeholder communication and a low level of integration of sustainability issues with business units. Environmental and social-oriented risk management (e.g., health care at the workplace) score high, which is almost always due to the strict regulations.

8.2.3 *Implementation*

8.2.3.1 Stakeholder Management

Stakeholder relations are not as developed in Hungary as on international average and are mostly managed more on a case-specific basis rather than in general. Informing stakeholders is most frequently used on a case-specific basis (56.0 %) and only one-third of the respondents use it in general, whereas the latter applies to the majority of the international sample (77.1 %). Dialogue with stakeholders is also used occasionally (60.0 %), while some companies (20.0 %) do not use this measure at all. Stakeholder involvement in decision-making processes is also low. Sustainability tools such as empowerment and delegating decision-making authority are not widespread, neither in other countries nor in Hungary.

8.2.3.2 Sustainability Management Tools and Standards

Sustainability tools are widely known in the country. Even tools less well-known in the international sample, such as material flow cost accounting (82.1 % in the Hungarian sample), environmental cost accounting (85.7 %) eco-indicators (71.2 %), sustainability control (82.1 %) and cross-impact analysis (71.4 %) were known by a majority of the Hungarian companies. Corporate citizenship and stakeholder dialogue were the only tools for which the knowledge of Hungarian companies was significantly lower than in the international sample. While 72.4 % know corporate citizenship and 75.2 % know stakeholder dialogue in the international sample, the numbers are 57.1 % for both tools in Hungary. This finding accords with responses given to other questions, which also indicate a low level of bilateral stakeholder communication.

The picture is relatively mixed regarding the application of sustainability management tools. Hungarian companies outperform the international sample for certain tools, e.g. environmental information systems (89.3 % compared to 52.1 % in the international sample), quality management systems (QMS) (92.9 and 75.6 %), eco-audits (75.0 % as compared to 51.7 %) and eco-indicators (75.0 and 50.9 %). This is partly due to the high penetration rate of the ISO 14001 standard among Hungarian companies. For example, ISO14001 requires the application of tools like eco-audits, environmental policy or mission statements, environmental indicators, environmental management system, etc. In Hungary the prevalence of ISO 14001 already exceeded 1,000 certifications in 2006, while there are only 28 companies registered for EMAS according to the database of Ministry of Rural Development (2013) today. Especially right after ISO 14001 was issued in 1996, Hungarian companies were pioneers in the implementation of this standard as measured by GDP-corrected certification numbers (Tóth 1999). This enthusiasm can be ascribed to two different factors. First, Hungarian companies are mostly in supplier positions and were obliged to comply with the requirements of their business partners.

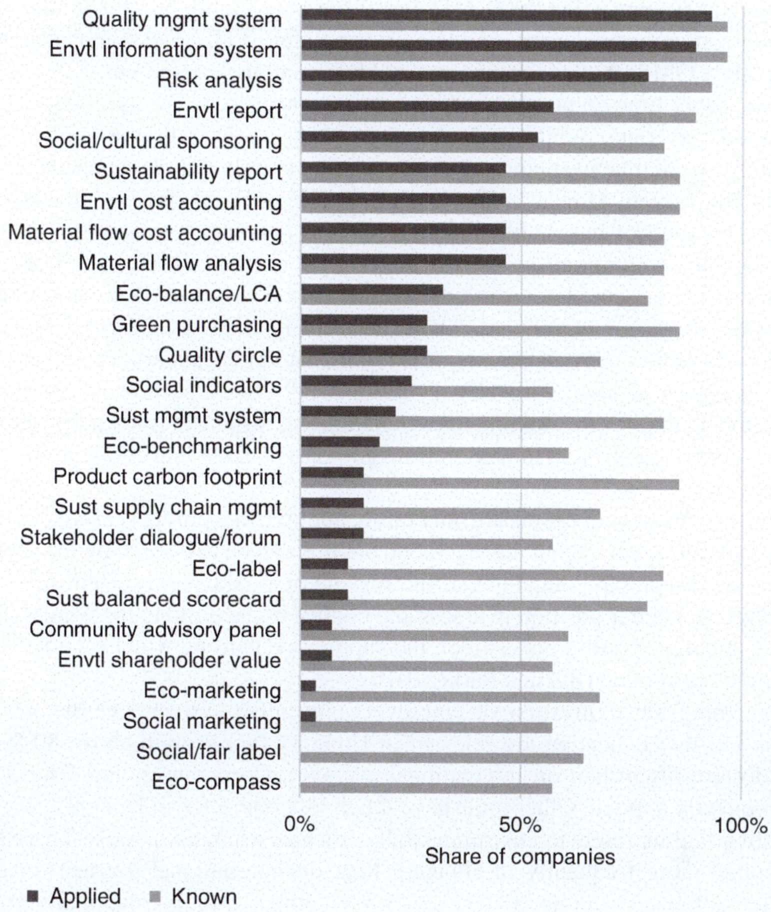

Fig. 8.1 Application of selected sustainability management tools in Hungary (per cent; $n_{HUN} = 28$)

Secondly, they risked losing their market position in case of non-implementation. Consultancy companies also played a major role in urging the implementation of certified management systems. Promotion of the ISO 14001 system occurred along with the quality assurance system of the ISO 9000 series, which was already widely applied and used in the country. The number of certified environmental management systems is especially high when compared to the size of the country (Tóth 2003). These standards require the use of numerous environmental management tools, internal and external audits, environmental programmes, training, etc.

The application of different sustainability management tools (see Fig. 8.1) is highly country and industry-specific. Companies in the chemical industry usually have more sophisticated sustainability management tools than their peers in less polluting industries (Széchy 2012). In Hungary, quality control and environmental management systems (ISO 9000, ISO 14001 etc.) are highly promoted, mainly by multinational companies and consultancies, and almost all large companies in the survey apply QMS (92.9 %), environmental information systems, etc.

At the same time some sustainability management tools are used less than on international average. While environmental management tools are applied more frequently, social management tools and integrated sustainability management tools are less frequent in Hungary. The country has significant deficits in using, for example, the tools of stakeholder dialogue (14.3 % in Hungary versus 45.9 % in the whole sample), community advisory boards (7.1 % versus 23.1 %), human resource control (35.7 % versus 51.4 %) and social audits (14.3 % versus 28.0 %). Corporate citizenship also has a significant deficit in the Hungarian sample (14.3 % versus 46.7 %). These findings can be interpreted such that large Hungarian companies are more engaged in dealing with global sustainability issues, such as environmental management, than with tackling local challenges, which is shown by the fact that, as mentioned above, Hungarian firms are only weakly embedded in the local community.

Some more innovative tools are only rarely applied, like product carbon footprint and sustainable supply chain management, but this can be explained by the supplier position of Hungarian companies in the big international supply chain networks. Buyers are in a better position to assess and control impacts along the whole supply chain, sometimes putting pressure on the suppliers regarding which standards and tools are to be applied (Bai and Sarkis 2010).

Concerning the motivation of employees, incentive system, flexible working time and further education are relevant in Hungary (application above 80 % and applied more often than on international average). Corporate giving (78.6 %) is also frequently applied. Other tools have much less relevance in Hungary. Another interesting finding refers to environmentally-oriented accounting tools. These tools are applied more frequently in Hungary than on international average: material and energy flow accounting: 53.6 % (+25.7 % compared to international average), material flow cost accounting: 46.4 % (+19.3 %), environmental cost accounting: 46.4 % (+23.4 %) and material flow analysis (46.4 and 31.5 %).

Among international standards and guidelines, ISO 14001 is dominant in Hungary (see Fig. 8.2). It is applied by almost 90 % of the surveyed companies. According to the ownership structure of certified companies, American, English and Japanese companies prefer ISO 14001 to other standards and guidelines, while EMAS is mainly promoted by German-owned companies (Ministry of Rural Development 2013). Also prominent is ISO 9000 (71.4 %), while OHSAS 18001/BS 8800 accounts for 39.3 % and the Global Reporting Initiative Guidelines (GRI) are known and applied by 35.7 %. Other standards such as OECD Guidelines, EFQM or UN Global Impact are less relevant in Hungary.

Almost all organisational units are rated below the international average for the question whether there is a need for developing suitable methods of environmental and social management. The Hungarian respondents are generally satisfied with

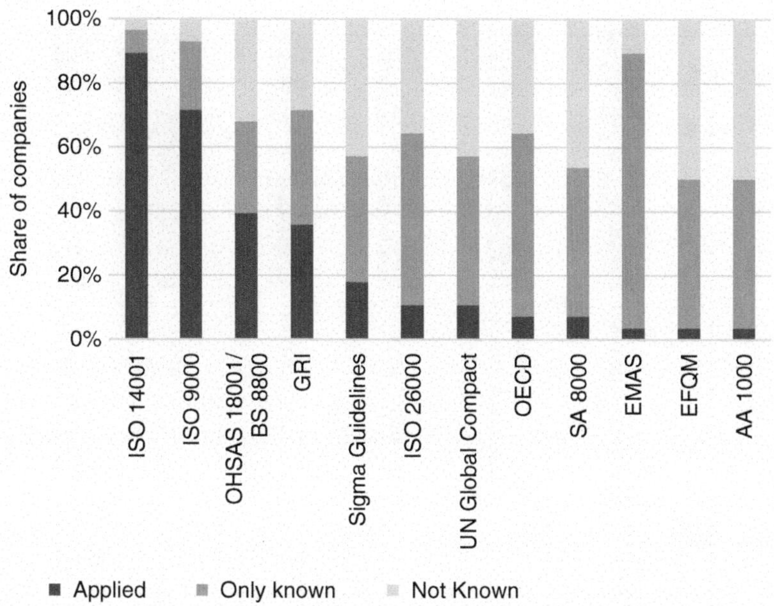

Fig. 8.2 Application of sustainability standards and guidelines in Hungary (per cent; $n_{HUN} = 28$)

the number and level of tools applied in the sustainability/CSR departments, which implies that they do not see the need for further development here. Again, environmental managers in Hungary usually consider they have a high level of expertise compared to their international peers.

8.2.3.3 Measurement

Impact measurement is highly connected to cost-cutting and regulations. It is thus very common that most forms of consumption and direct emissions are measured by a majority of the Hungarian companies (see Fig. 8.3). All environmental impacts surveyed, with the exception of biodiversity, are monitored by a higher share of companies than in the international sample. This goes for energy, material and water consumption, emissions and transport. Biodiversity monitoring is slightly below the international sample.

Transport is a more crucial problem since a company is not always responsible for measuring the emissions of their suppliers, so it is mostly not taken into consideration. However, 74.1 % of the Hungarian companies surveyed take account of their impact on transportation, which is still above the international average. Another reason why a quarter of the companies do not measure the development of

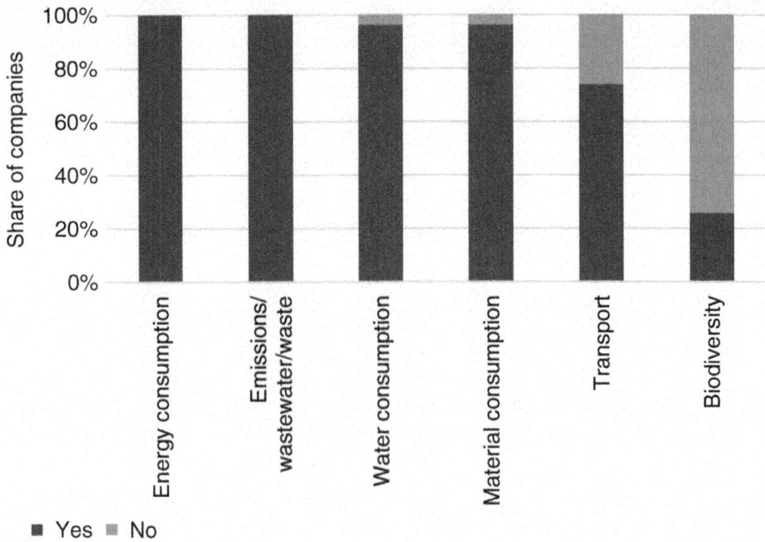

Fig. 8.3 Measuring impacts of environmental issues in Hungary (per cent; $n_{HUN} = 27$, n_{INT} ranging from 442 to 454)

transport can simply be that they are not transporting anything. So, the higher share of manufacturing companies in the Hungarian sample explains the high value of its measurement. Biodiversity measurement is carried out by only 25.9 % of Hungarian companies, which reflects that this measurement is highly debated and that there is a lack of suitable indicators.

The measurement of the social issues workplace/employment, occupational health and safety as well as training and development is usually necessary in order to comply with regulations. However, aspects such as consumer protection or child labour, forced and compulsory labour are neglected as social indicators; only around one-third of all companies pays attention to these issues. Measurement of diversity and equal opportunity accounts for 57.7 % among the Hungarian companies surveyed. Freedom of association/right to collective bargaining is considered to be strong in Hungary (a result that is also highly culture dependent) and accordingly has a significantly higher measurement (65.4 %) than in other countries surveyed.

Most environmental issues are reported by the majority of companies to have improved over the 2 years preceding the survey, except for biodiversity loss and transport. Changes in the impacts of social issues are reported as small or are not measured by a relatively large share of companies. Unfortunately, the measurement of the impacts of sustainability management on company success or competitive advantage is undertaken by only 12–46.2 % of the companies surveyed (see Fig. 8.4), a considerably smaller share than in the international sample. The overall impact of environmental management is given a positive evaluation, but the impact of social management on company success is still evaluated less positively and is viewed sceptically.

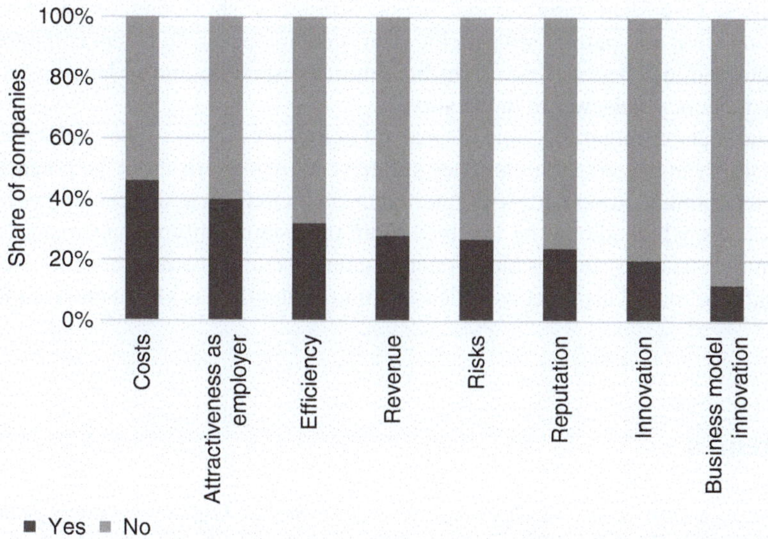

Fig. 8.4 Measuring the impact of sustainability management on company success (per cent; n_{HUN} ranging from 25 to 26)

8.3 Conclusion and Outlook

Although the Hungarian companies surveyed represent only a small segment of the Hungarian economy, certain tendencies and characteristics can still be identified. Social and environmental policies are most often handled separately and are not integrated into the core business but instead into some segments and organisational units of the companies.

Sustainability management tools mandated by the ISO 14001 standard, including audits, are applied frequently, as the penetration level of ISO 14001 certification is very high in the country. The application of some tools, not surprisingly, lags behind the international average. This is especially true for tools requiring stakeholder participation and dialogue as well as tools relating to the local rather than to the global environment. Environmental management tools are better known and more often applied than social management tools in the country. Due to the dominance of engineering experts in sustainability management, sustainability management tools tend to be related to environmental or material flow cost accounting and less to social issues.

Hungarian companies report improvements in energy, water and material consumption more often than their international peers. This is due to the fact that energy, material and water savings are often perceived as economic issues and issues with a financial return. Economic and environmental objectives converge regarding these issues. Hungarian companies feel less pressure to manage sustainability issues from external stakeholders, especially from consumers, rating agencies and national authorities, than their international peers. Still, they show deficits in two-way communication and participative methods in stakeholder management compared to their international peers. This might be culturally specific and is lagging behind general tendencies elsewhere in the survey.

In summary, Hungarian companies score high with regard to the application of sustainability management tools. They still have more to learn about the potential of bilateral communication both with internal decision makers and especially external stakeholders, which should be involved more in sustainability management. Local environmental issues, such as biodiversity, human resource management or diversity of workforce should also receive at least as much attention as global environmental issues.

References

Bai C, Sarkis J (2010) Green supplier development: analytical evaluation using rough set theory. J Clean Prod 18(12):1200–1210

Boda G, Kállay L, Stocker M (2010) Versenyszféra teljesítményéről nemzetközi összehasonlításban TM 102. sz. Műhelytanulmány BCE Vállatgazdaságtan Intézet Budapest 2010. http://edok.lib.uni-corvinus.hu/465/1/TM102_Boda__Kallay_Stocker.pdf. Accessed 10 Nov 2013

Bodorkós B, Pataki G (2009) Local communities empowered to plan? Action Res 7(3):313–334

Csutora M, Harangozó G (2009) European Sustainability Reporting Association report for Hungary, 2008. Műhelytanulmány. ESRA, London. http://www.sustainablereporting.eu/hungary. Accessed 09 Sept 2013

Csutora M, Kerekes S, Harangozó G, Németh P, Zsóka A (2006) Befolyásunk a vállalatok környezetvédelmi magatartására. In: Kerekes S (ed) Környezeti Nézőpontok. Budapesti Corvinus Egyetem, Budapest

Gelei A (2003) Competitiveness: a match between value drivers and competencies in the Hungarian automotive supply chain. IFPSM Summer School Publications. http://ifpsm.org/index.php/pub/summer-school-2003-2004-publications/. Accessed 09 Nov 2013

Harangozó G, Kerekes S, Zsóka Á (2010) Environmental management practices in the manufacturing sector? Hungarian features in international comparison. J East Eur Manag Stud 4:312–347

Kerekes S, Harangozó G, Zsóka Á, Németh P (2003) Environmental policy tools and firm-level management practices, OECD National Report: Hungary. Kutatási jelentés. BCE, Környezetgazdaságtani és Technológiai Tanszék, Budapest

Kerekes S, Wetzker K, Csutora M, Harangozó G, Szőnyi L (2006) Electrolux Lehel – more than a profitable enterprise Budapest-Jászberény. http://www.environmentalmanager.org/wp-content/uploads/2008/04/electroluxcorvinuseangol1.pdf. Accessed 10 Nov 2013

Klauber M (2008) A járműipari ágazati stratégia kialakítását megalapozó szakmai átvilágító tanulmány. KOPINT-TÁRKI, Budapest

KSH (Hungarian Statistical Office) (2013) Magyar Statisztikai Évkönyv 2012. Budapest

Magerholm Fet A, Zilahy G (2011) A környezetbarát beszerzés gyakorlata Norvégiában és Magyarországon. In: Csutora M, Hofmeister TÁ (eds) Fenntartható fogyasztás? A fenntartható fogyasztás gazdasági kérdései. Budapesti Corvinus Egyetem, Budapest, pp 110–124

Ministry of Rural Development (2013) EMAS register. http://emas.kvvm.hu/company.php?l=. Accessed 20 July 2013

Nagy B (2005) Nők a vezetésben. In: Nagy I, Pongrácz T, Tóth IG (eds) Szerepváltozások. Jelentés a nők és férfiak helyzetéről. TÁRKI, Ifjúsági, Családügyi, Szociális és Esélyegyenlőségi Minisztérium, Budapest, pp 44–56

Schaltegger S, Lüdeke-Freund F, Hansen E (2012) Business cases for sustainability. The role of business model innovation for corporate sustainability. Int J Innov Sustainable Dev 6(2):95–119

Schaltegger S, Harms D, Hörisch J, Windolph SE, Burritt R, Carter A, Truran S, Crutzen N, Ben Rhouma A, Csutora M, Tabi A, Kokubu K, Kitada H, Haider MB, Kim J-D, Lee K-H, Moneva JM, Ortas E, Álvarez-Etxeberria I, Daub C-H, Schmidt J, Herzig C, Morelli J (2013) International corporate sustainability barometer: a comparative study of 11 countries. Centre for Sustainability Management, Lüneburg

Szechy A (2012) Determinants of environmental innovation in the Hungarian chemicals sector. Interdiscip Environ Rev 13(2/3):139–157

Tóth G (1999) ISO 14001 speedometer – offers fresh look at the certification derby. Bus Environ X(8):3–5

Tóth G (2003) Evaluation of environmental performance of companies. Soc Economy 25(3): 383–402

Chapter 9
Corporate Sustainability Barometer in Japan

Katsuhiko Kokubu, Hirotsugu Kitada, and M. Badrul Haider

Abstract This chapter describes the state of the art of corporate sustainability practice in Japan in comparison with the international sample. It is based on a questionnaire survey conducted as part of the International Corporate Sustainability Barometer 2012 (ICSB) Project. In many instances Japanese scores are found to be above the international average, which indicates the significance of sustainability among the investigated companies. While a number of similarities are observed with regard to the intention, integration and implementation of corporate sustainability, some particularities specific to the Japanese context are also noted. For example, communities, consumers and trade associations are seen to be more promoting in Japan than in the international sample. Companies in Japan are also found to prefer different methods and guidelines for their sustainability management. Finally, the findings are explained based on the contextual factors in Japan.

9.1 Introduction

The global call for corporate sustainability is now more urgent than ever. Companies today face an unprecedented degree of critical scrutiny and rising expectations from a wide range of stakeholders (Zadek 2007). Corporate executives need to

---

K. Kokubu (✉)
Graduate School of Business Administration, Kobe University, Kobe, Japan
e-mail: kokubu@kobe-u.ac.jp

H. Kitada
Faculty of Business Administration, Hosei University, Tokyo, Japan
e-mail: kitada@hosei.ac.jp

M.B. Haider
Institute of Business and Accounting, Kwansei Gakuin University, Kobe, Japan
e-mail: haider@people.kobe-u.ac.jp

S. Schaltegger et al. (eds.), *Corporate Sustainability in International Comparison*, Eco-Efficiency in Industry and Science 31, DOI 10.1007/978-3-319-06227-3_9,
© Springer International Publishing Switzerland 2014

find new ways to address the social, economic and environmental effects of doing business while balancing the conflicting demands on their attention, time and company resources (Schaltegger and Burritt 2005; Epstein 2008). While historically the concept of sustainability is deeply rooted in Japanese business practices, western-style corporate social responsibility (CSR) practice first emerged in the 1980s and gained much recognition in the 1990s (Fukukawa and Moon 2004; Kawamura 2005a, b). The objective of this chapter is to present the state of the art of corporate sustainability practice in Japan compared with the international average as based on a questionnaire survey conducted as part of the International Corporate Sustainability Barometer 2012 (ICSB) Project coordinated by the Centre for Sustainability Management (CSM) in Lüneburg, Germany (see Schaltegger et al. 2013).

The chapter is structured as follows. First, the sustainability context in Japan is briefly explained. Then the characteristics of the sample are described. Key findings of the survey are analysed in Sect. 9.2 based on the three research foci of the ICSB Project: the intention, integration and implementation of sustainability management. Finally, the conclusion describes and highlights the particularities of the Japanese corporate sustainability practices.

9.1.1 Context of Sustainability in Japan

Historically sustainability is deeply rooted in Japanese business practices (Wokutch 1990; Lewin et al. 1995; Wokutch and Shepard 1999; Kawamura 2003; Fukukawa and Teramoto 2009). However, the current practice of corporate sustainability is largely influenced by the western concept of CSR brought to the country during the 1980s and 1990s (Fukukawa and Moon 2004). A number of interrelated domestic and global factors are identified as the key drivers of this new movement in Japan. These include increased social awareness about public health and the environment, government guidelines and laws, initiatives by business associations and the influence of globalisation (Kokubu et al. 2003; Fukukawa and Moon 2004; Kawamura 2005a).

Corporate practices are often affected by the cultural background of society (Burritt et al. 2003) and Japan is considered to be more group-oriented than western society (Hofstede 2001). Members in a group show primary loyalty to their group and feel a much greater sense of duty and responsibility to their group than to others in society. This in-group/out-group distinction in Japanese society effectively reduces tensions with the main stakeholders of the company and may exclude marginal or indirect stakeholders. For example, while Japanese companies were championed for their contribution to communities or employees, they were criticised on other issues such as equal employment opportunities for minorities, foreigners or women and the protection of the environment (Wokutch and Shepard 1999). Thus, a strategic integration has been observed between corporations and their key stakeholders, such as investors, employees, consumers, suppliers and communities (Tanimoto and Suzuki 2005). With the aim of rapid economic growth after World War II, corporate interests were given priority in such contexts. This kind of corporate control, according to Vilanova and Tanimoto (2009), has reduced stakeholder pressure for corporate sustainability.

Massive industrialisation and high economic growth in post-war Japan were accompanied by rampant environmental degradation frequently neglected by the companies. The terrible effects of environmental pollution were demonstrated by the widely reported public health problems of Yokkaichi asthma, Minamata disease or Itai-Itai diseases (Fukukawa and Moon 2004). Public confidence in companies was further diminished during the 1980s and 1990s after a series of business scandals ranging from unfair financing, illegal payoffs, mass food poisoning, misrepresentation of annual reports to the failure of financial institutions such as Yamaichi Securities and Hokkaido Takushoku Bank (Kawamura 2005b). This new consciousness in society "led corporations to need to announce their (re-)commitment to 'co-habitation' [kyosei] with society" (Fukukawa 2010).

The Japanese Government has had significant influence on the increased attention to corporate sustainability (Kokubu et al. 2003; Kokubu and Nashioka 2005; Choi and Aguilera 2009). During the last two decades government ministries such as the Ministry of Environment, Ministry of Economy, Trade and Industry and Ministry of Health, Labour and Welfare have issued a number of decrees and voluntary guidelines on corporate social and environmental responsibilities (Fukukawa and Moon 2004). As Japanese society is less legalistic than western countries, instead of specific rules and regulations the government in Japan usually uses voluntary guidelines and recommendations to ensure a cooperative relationship between companies and the government regarding CSR activities (Lewin et al. 1995). This approach gives greater flexibility to companies in interpreting and adopting government guidelines and thus contributes to creating a favourable business environment (Choi and Aguilera 2009). The active participation of businesses in developing guidelines also increases the chances of success in application (Choi and Aguilera 2009). Leading business associations such as Nippon Keidanren (Japan Federation of Economic Organizations) and Keizei Doyukai (Japan Association of Corporate Executives) also responded positively and announced charters and guidelines for CSR activities during the 1990s (Fukukawa and Moon 2004; Keidanren 2010).

The impact of globalisation has been another frequently cited reason for improved corporate sustainability in Japan (Tanimoto and Suzuki 2005; Fukukawa and Teramoto 2009). Japanese companies have extensive trading, manufacturing and other operations outside Japan, including with USA, UK and Australian companies, which often have more sophisticated CSR practices. This provides opportunities for Japanese companies to learn new CSR practices in those host countries and then to introduce these concepts on the domestic level. Recently, Japanese companies are receiving increased attention from eco-fund managers in the USA and Europe, which according to Kawamura (2003) is one of the key drivers for the current development of CSR in Japan.

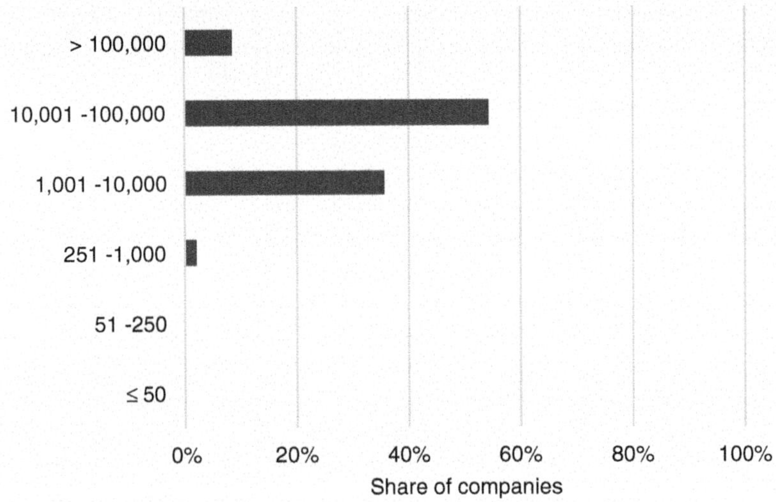

Fig. 9.1 Characteristics of the companies surveyed (number of employees; $n_{JAP} = 48$)

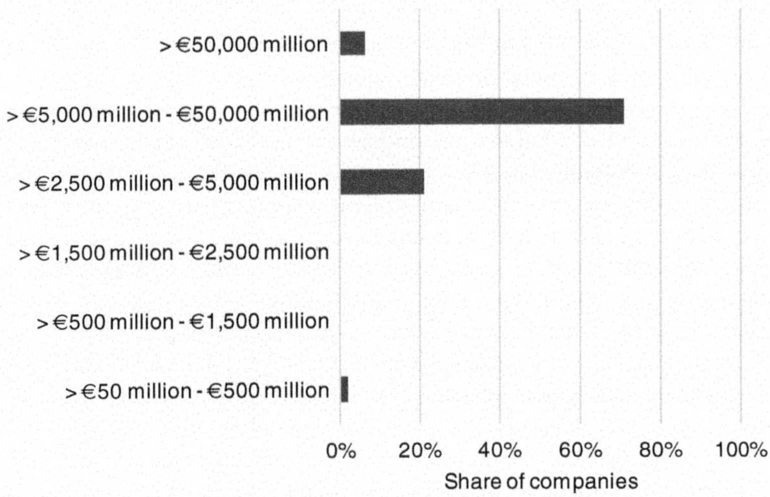

Fig. 9.2 Characteristics of the companies surveyed (sales revenue; $n_{JAP} = 48$)

9.1.2 The Japanese Sample

A questionnaire was sent to the top 300 companies based on annual revenues and a total of 48 valid questionnaires were received (a response rate of 16.0 %). The majority of the sample consists of relatively large companies both in terms of number of employees and sales revenues (see Figs. 9.1 and 9.2). While about

90 % of the sample has employees between 1,001 and 100,000, about 71 % of the companies have sales revenues within the range of more than €5,000–€50,000 million. Although in terms of sales the sample shows higher values than the international sample, with respect to the number of employees they are comparable. The investigated Japanese companies also have a strong international presence in term of sales, with 40.5 % of the sample having non-domestic sales of 40–80 %. The sample is also diverse with regard to industry, with about one-third coming from industry, capital goods & building as well as commodities, auxiliary material, energy, chemical & pharmaceutical industry each. While in the international sample, finance & services industry dominate, the Japanese sample only contains 12.5 % from these industries.

9.2 Analysis

The findings of the survey are analysed based on the three research foci of the ICSB: intention, integration and implementation of sustainability management.

9.2.1 Intention

9.2.1.1 Motivation

The motivation to engage in corporate sustainability may come from within the organisation and from the external environment (Epstein 2008). Once engaged, corporations usually expect business value from the implementation of sustainability strategies. However, a number of factors may also deter companies from engaging in sustainability practices. This section highlights the stakeholders' influence, key drivers and obstacles, impacts and issues in sustainability management.

Stakeholders today are much more concerned about the social and environmental impacts of companies and provide an incentive to engage in sustainability practices. Therefore, a question was asked about stakeholder impact on the implementation of corporate sustainability and the results are shown in Fig. 9.3. Most external stakeholders are evaluated as being more promoting for the implementation of corporate sustainability in Japan than in the international sample. The most promoting external stakeholders in Japan are community, consumers (end users), trade associations, NGOs and rating agencies. It is noteworthy that consumers and trade associations, which have a relatively low ranking in the international sample, are influential in Japan. As consumer awareness of sustainability has increased, companies in Japan, especially CSR departments, want to focus on consumers as the core of their activities. A survey by METI (2004) also reveals that more than 60 % of the respondents in Japan prefer to buy from companies that are socially responsible and have a sound ethical policy. Trade associations, including representative bodies such as Keidanren (Japan Federation of Economic Organizations) and Keizei Doyukai

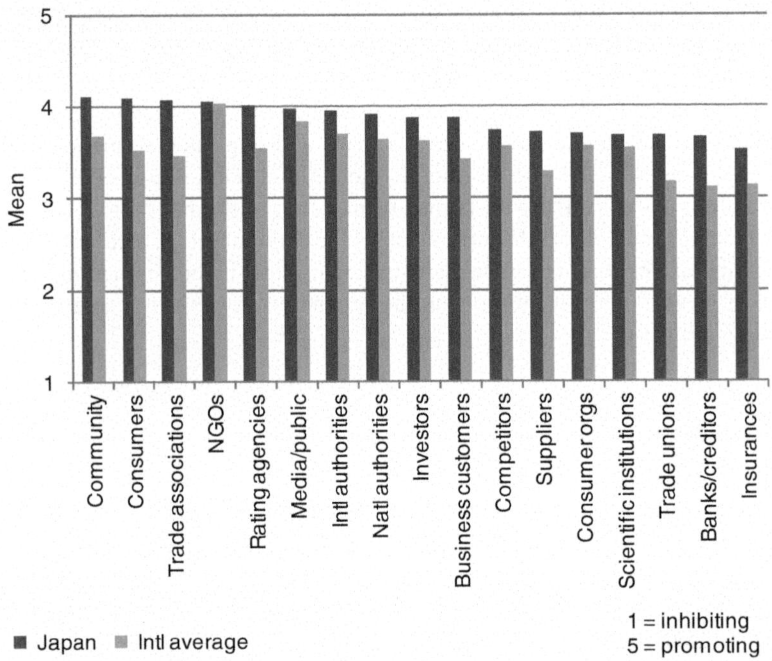

Fig. 9.3 Stakeholders' impact on the implementation of corporate sustainability (n_{JPN} ranging from 39 to 46, n_{INT} ranging from 393 to 450)

(the Japan Association of Corporate Executives), are also actively promoting sustainability through different voluntary guidelines, charters and industry specific conventions (Keidanren 2003; Keizai Doyukai 2004).

The responses about the impact of factors that promote and inhibit corporate sustainability management are given in Figs. 9.4 and 9.5 respectively. As in other countries, commitment by top management, corporate philosophy and legislation (statutory rules and guidelines) are considered to be the three most important drivers for the implementation of corporate sustainability in Japan (see Fig. 9.4). However, the Japanese values for these three factors are even higher than on international average.

On the other hand, with slightly higher values in Japan than on international average, legislation (bureaucracy/legal regulations), methodological complexity of sustainability management and lack of know-how in the company are cited as the most important barriers to implementing sustainability management. This is a sharp contrast to the international sample, where lack of personnel capacities, lack of financial capacities and lack of government incentives (subsidies/tax advantages) are considered to be the most important barriers to corporate sustainability (see Fig. 9.5).

Another question served to understand the positive impacts of sustainability management implementation. Similar to their international counterparts, the Japanese corporations mostly expect employee motivation, enhancing and safeguarding

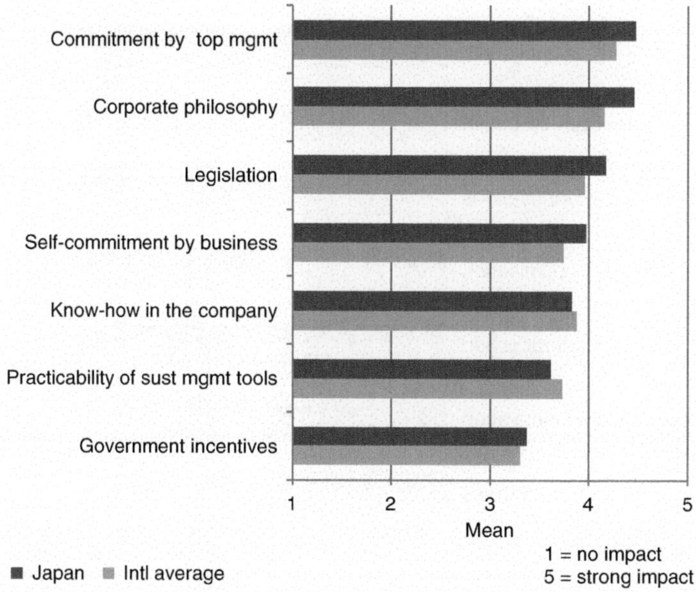

Fig. 9.4 Promoting factors for implementation of corporate sustainability (n_{JPN} ranging from 42 to 47, n_{INT} ranging from 446 to 460)

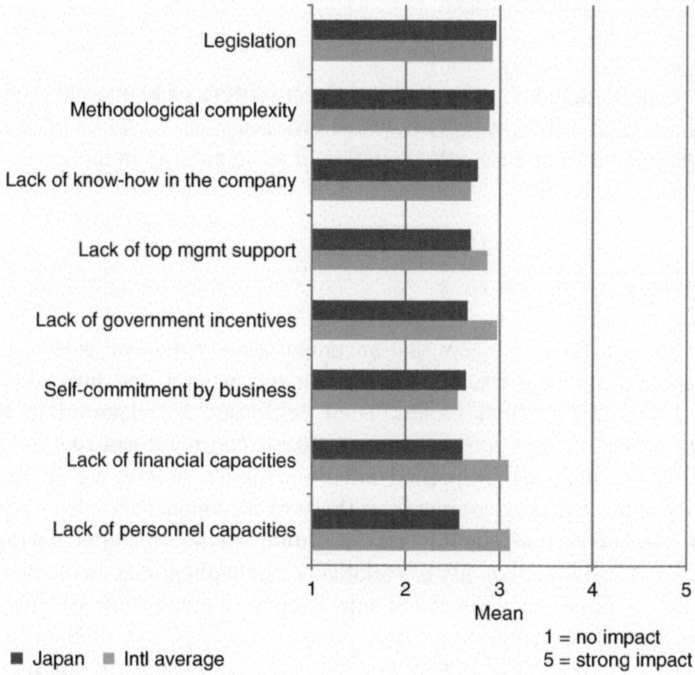

Fig. 9.5 Inhibiting factors for implementation of corporate sustainability (n_{JPN} ranging from 44 to 47, n_{INT} ranging from 438 to 449)

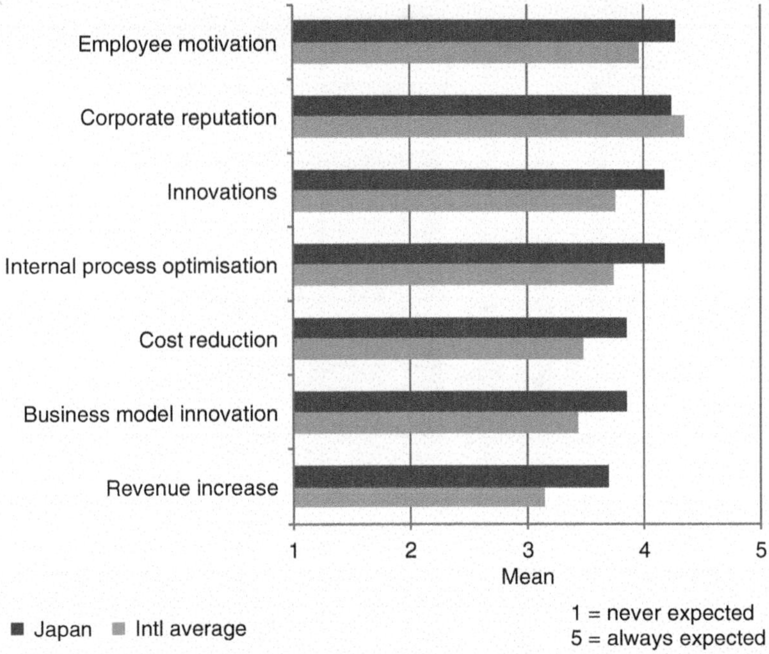

Fig. 9.6 Positive impacts from the implementation of corporate sustainability (n_{JPN} ranging from 46 to 48, n_{INT} ranging from 450 to 458)

corporate reputation, innovation (products, processes etc.) and internal process optimisation (e.g. more efficient organisational processes and methods of production) as positive impacts from the implementation of sustainability in their organisations (see Fig. 9.6).

9.2.1.2 Issues

Sustainability is a broad concept that encompasses a variety of aspects including social and environmental issues. The respondents were asked how closely they managed a number of such issues. Their responses are reported in Fig. 9.7. Around the world, energy consumption, emissions/wastewater/waste and material consumption are the most important environmental concerns for companies. A particularity of the Japanese companies is that they also manage biodiversity closely, a topic that receives considerably less attention on international average. This increased attention to biodiversity is a relatively recent phenomenon that was mainly triggered by the Japanese government's decision to host the COP 10 – Conference of the Parties to the Convention on Biological Diversity (CBD) in Nagoya in 2010 (Biodiversity Network Japan 2008). Interest is also influenced by the biodiversity

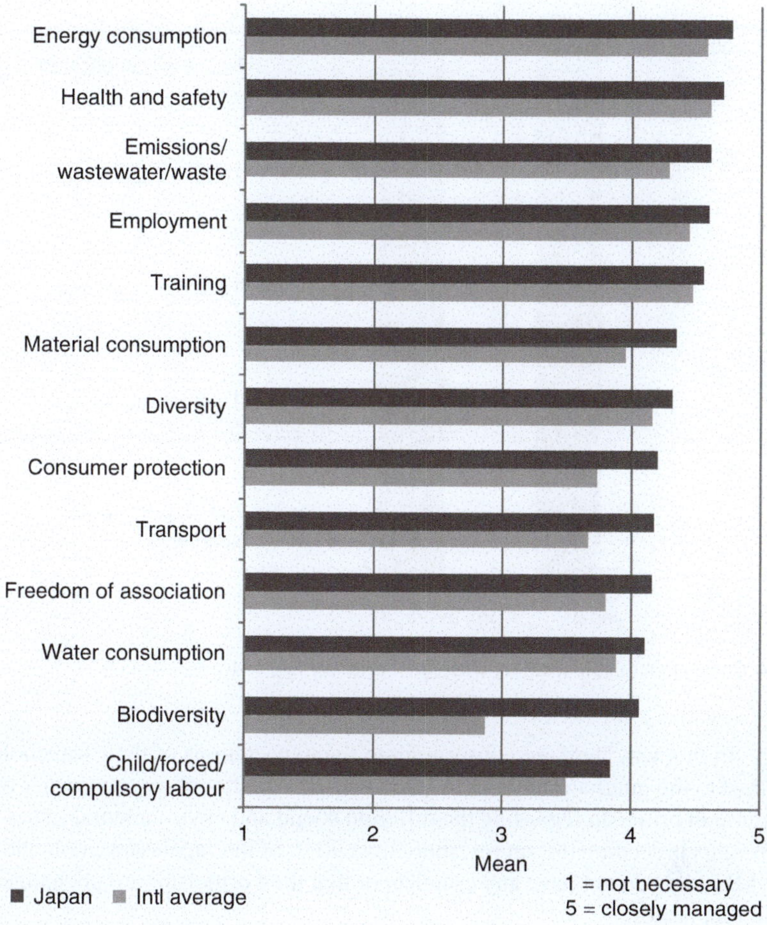

Fig. 9.7 Environmental and social issues managed by companies (n_{JPN} ranging from 44 to 48, n_{INT} ranging from 442 to 463)

guidelines for the private sector included in the Third National Biodiversity Strategy of Japan adopted by the Cabinet in 2007 (Ministry of the Environment of Japan 2008). Nippon Keidanren has also been playing a pivotal role through the Nippon Keidanren Committee on Nature Conservation and the Keidanren Nature Conservation Fund (KNCF), which were created in 1992 (Nippon Keidanren Committee on Nature Conservation 2008). Furthermore, Japan supported the development of the Business and Biodiversity Initiative (Schaltegger and Beständig 2010).

Among the social issues, management of occupational health and safety, workplace/employment as well as training and development are given the highest priority by companies in Japan as well as on international average. The respondents believe that these are also the most important issues that their stakeholders want them to

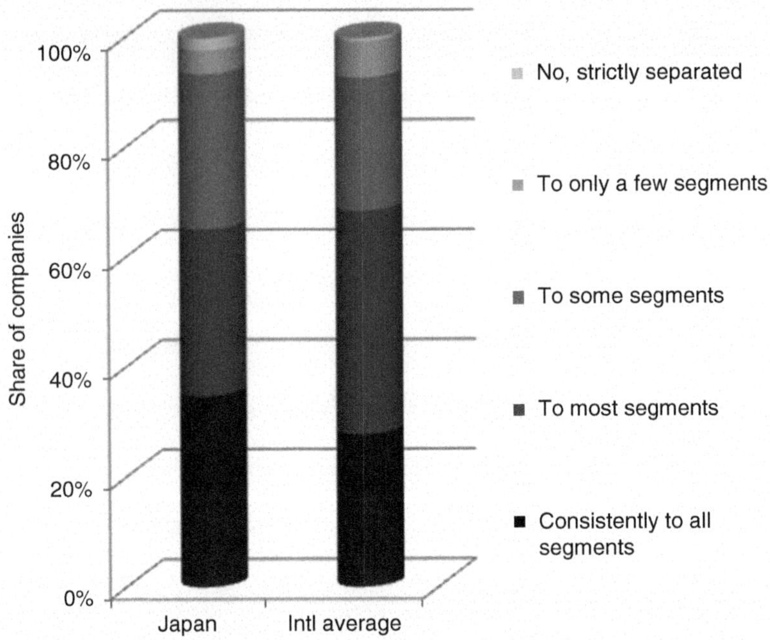

Fig. 9.8 Connection of sustainability with core business ($n_{JPN} = 46$, $n_{INT} = 457$)

manage. In this way they are conforming to the expectations of their stakeholders. In contrast to the international sample, where up to a quarter of the companies report an increase in criticism regarding the different social and environmental issues over the two years preceding the survey, more than 90 % of the Japanese sample indicate that stakeholders do not have any criticism or that their criticisms are unchanged for the different issues.

9.2.2 Integration

9.2.2.1 Connection to Core Business

'Sustainability' is no longer only a buzzword. It is a strategic business issue and for success in the long run management must integrate sustainability into their core business models (Epstein 2008; Schaltegger et al. 2012). This section focuses on the integration of sustainability with core business activities and with all organisational units. Figure 9.8 shows that about two-thirds of the surveyed companies are either consistently connecting sustainability to all segments of their core businesses (34.8 %) or most segments of the core business (30.4 %). The practice is similar to the international average, where also about 69 % of the companies are either

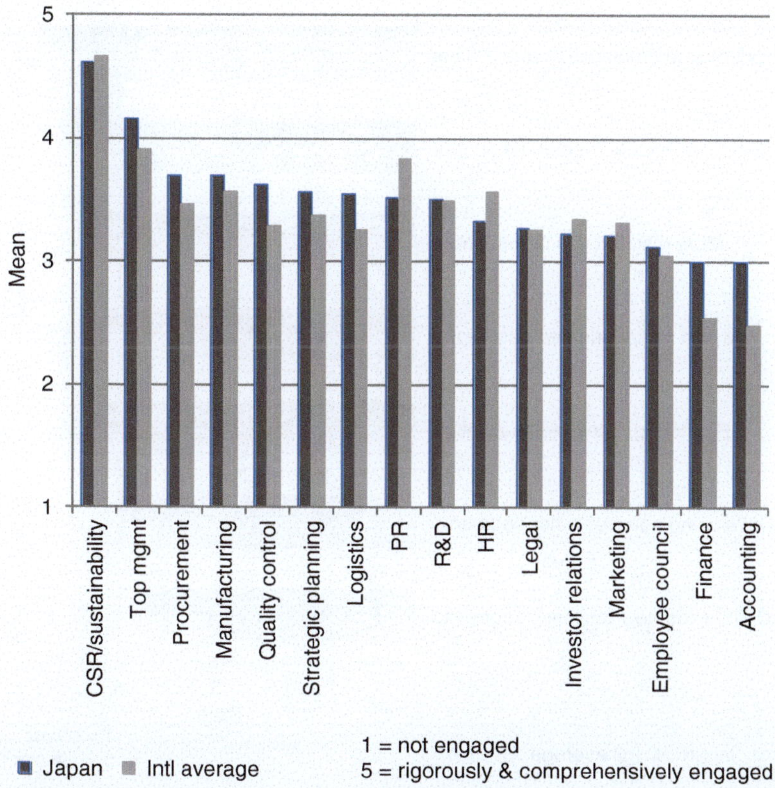

Fig. 9.9 Organisational units engaging in sustainability measures (n_{JPN} ranging from 35 to 44, n_{INT} ranging from 312 to 337)

consistently connecting sustainability to all segments of the core business (27.8 %) or most segments of the core business (40.7 %).

9.2.2.2 Involvement of Organisational Units

Integrating sustainability requires organisation-wide efforts. As a consequence, different functional units of the organisation are affected as corporations increase their awareness of sustainability principles (Epstein 2008). The ICSB survey included a question regarding the engagement of organisational units for sustainability measures (see Fig. 9.9). CSR/sustainability (including EHS/environment/health/occupational safety) is the most crucial organisational unit in sustainability management. By setting the tone for sustainability in an organisation, top management also plays an important role. In Japan, procurement/purchasing and manufacturing departments are also actively involved in corporate sustainability measures. Public

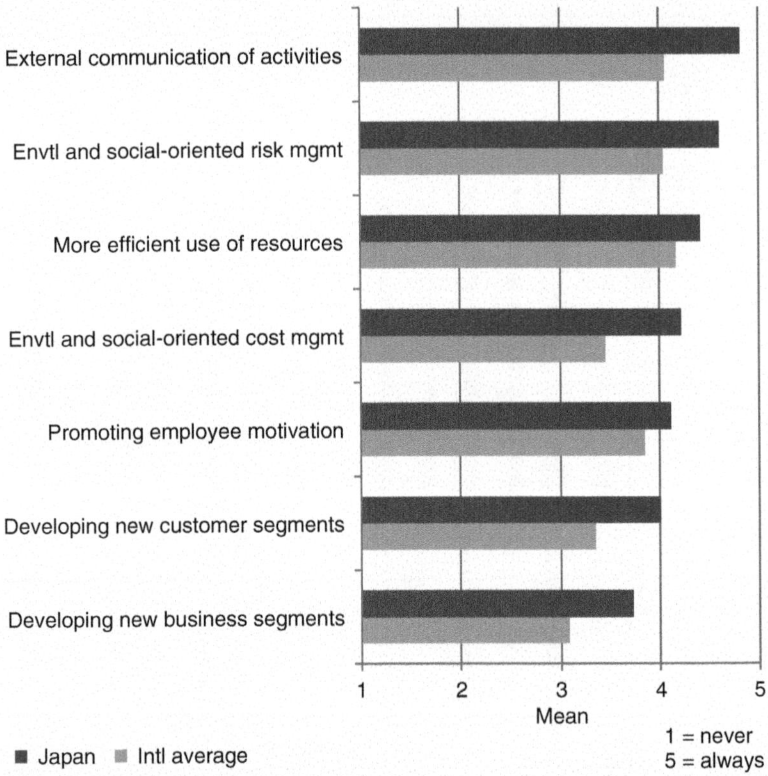

Fig. 9.10 Implementation of sustainability measures (n_{JPN} ranging from 45 to 47, n_{INT} ranging from 397 to 405)

relations/corporate communication and the personnel department/HR, which are seen to be important on international average, however, are relatively less involved in Japan.

9.2.2.3 Business Case Drivers for Sustainability

The respondents were asked whether a number of sustainability measures were implemented or not (see Fig. 9.10). The communication of social and environmental activities to external stakeholders (e.g. sustainability reporting) is the key sustainability measure that most organisations in Japan have implemented. Other important activities include environmental and social-oriented risk management (e.g. health care at the workplace) and producing with more efficient use of resources (e.g. optimising production processes). Although the companies in the international sample have ranked producing with more efficient use of resources as their top priority, they also agreed that these are the three most important measures of sustainability implemented by them. Yet, the values are higher in Japan.

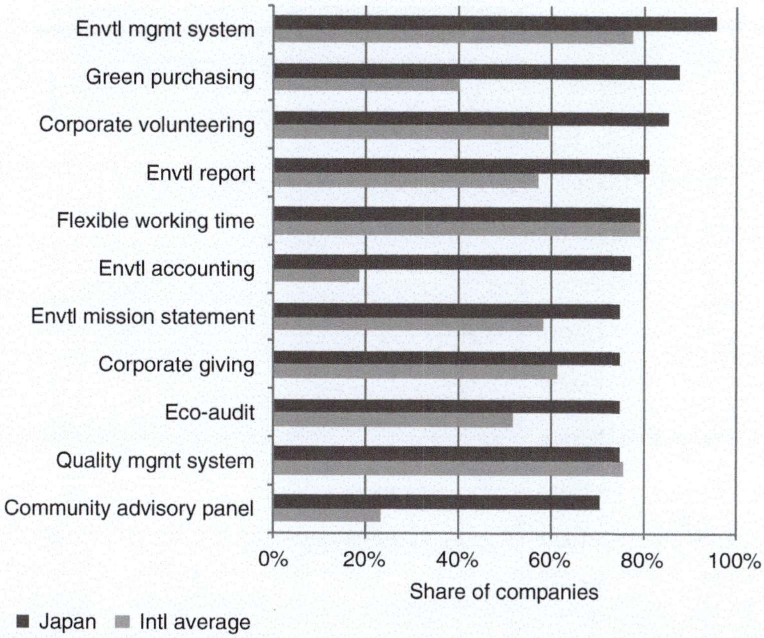

Fig. 9.11 Broadly applied sustainability management tools (per cent; $n_{JAP} = 48$)

9.2.3 *Implementation*

The successful implementation of corporate sustainability largely depends on management's ability to design and align with various sustainability measurement and management systems, programmes and tools (Epstein 2008). These will help companies to systematically identify, measure and appropriately manage their sustainability responsibilities and risks. Understanding and managing stakeholder expectations and activities are also important for the successful implementation of the corporate sustainability agenda. Therefore, the final section discusses such tools and guidelines or standards that are used in sustainability and stakeholder management.

9.2.3.1 Sustainability Management Tools

Figure 9.11 shows which sustainability management tools are widely known and applied by Japanese companies. Of the respondents 70 % or more note that these tools are known and applied in their companies. Among these tools, in comparison to the international sample, Japan is far ahead in utilising environmental accounting and reporting, green purchasing, community advisory panel, eco-audit

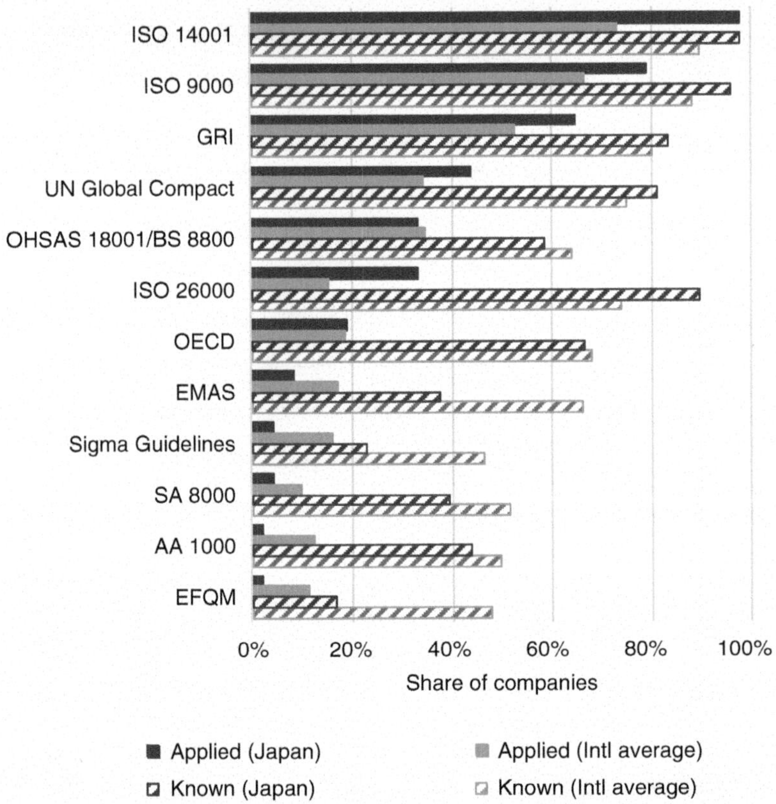

Fig. 9.12 Standards/guidelines known and applied in the companies (per cent; $n_{JAP} = 48$)

and corporate/employee volunteering. Employee-oriented tools like human resource control, incentive system and further education, which are used by more than 50 % of the international sample, however, are less known and used in Japan (not shown in Fig. 9.11).

As a number of voluntary standards, codes and principles are available for sustainability management, companies must decide which are most appropriate for their business strategies. Japanese companies' adherence to the ISO standards is clearly demonstrated in the survey (see Fig. 9.12). Almost all the investigated companies have adopted ISO 14001 and nearly 80 % of them apply ISO 9001. The Global Reporting Initiative Guidelines (GRI) and UN Global Compact are the other two widely known and used standards/guidelines in the Japanese context. Whereas internationally nearly half of the respondents at least know EFQM (including S-EFQM), EMAS and Sigma Guidelines, these standards and guidelines are not common in Japan.

The respondents were also asked about the methods of stakeholder management and the results are shown in Fig. 9.13. Informing stakeholders through websites

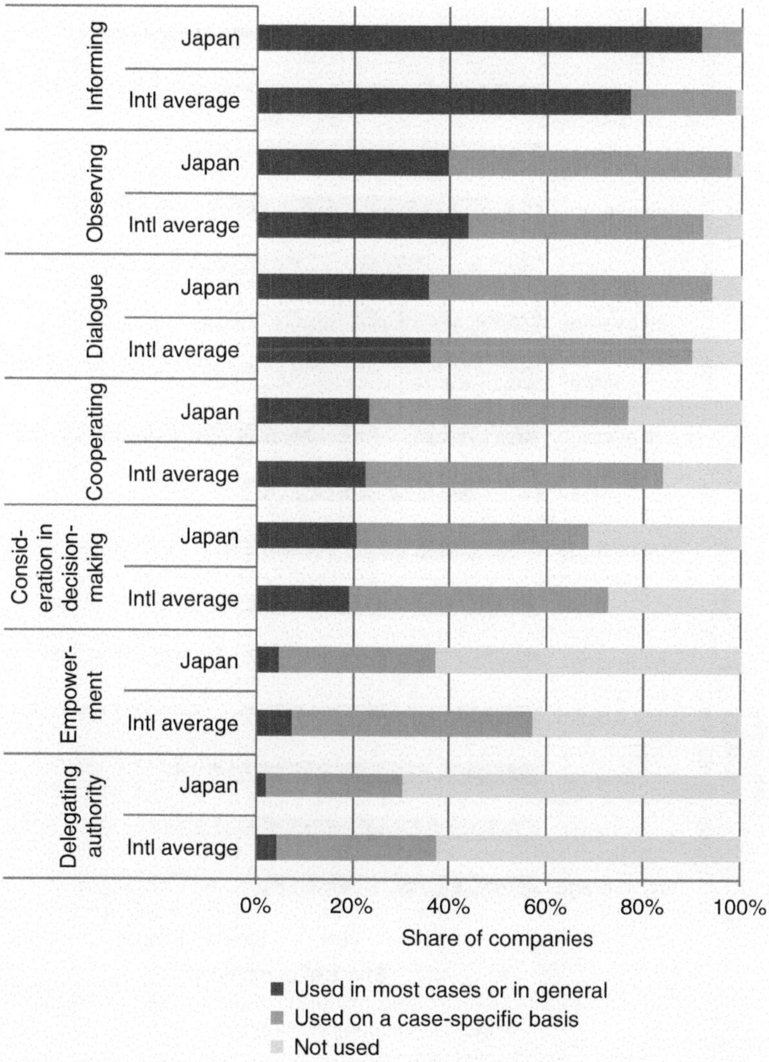

Fig. 9.13 Methods of stakeholder management (n_{JPN} ranging from 43 to 48, n_{INT} ranging from 438 to 458)

or the press is the most important medium for stakeholder management, as more than 90 % of the Japanese respondents indicate that they use this technique in most cases or in general. Furthermore, a good percentage of the companies (35.4 % in most cases and 58.3 % on a case basis) are using stakeholder dialogues (e.g. questionnaires, dialogue forums), whereas, similar to the international sample, empowerment of stakeholders (e.g. providing financial support) and delegating decision-making authority to the stakeholders are still rare in Japan.

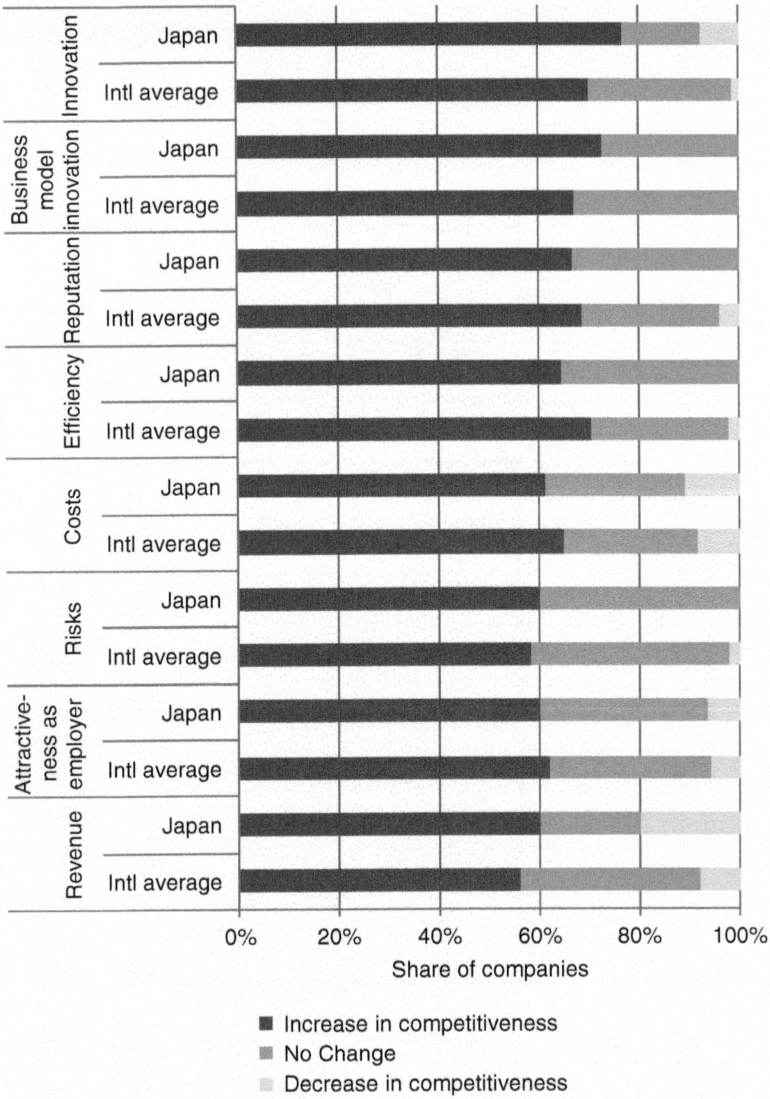

Fig. 9.14 Impact of sustainability management on corporate competitiveness (n_{JPN} ranging from 10 to 18, n_{INT} ranging from 119 to 206)

9.2.3.2 Measurement

The majority of the companies in Japan stated that their environmental and social management have a significant positive impact on the overall success of their organisation. To analyse how these positive impacts are generated, the questionnaire surveyed the specific drivers of business cases for sustainability the companies use.

The results are shown in Fig. 9.14. Sustainability is a source of innovation in the form of products, processes or business models. Enhanced competitiveness in terms of brand reputation, cost minimisation and internal efficiency is also important. However, one-fifth of the respondents also note that sustainability management decreases their competitiveness with respect to revenues/sales/profits.

9.3 Conclusion

This chapter describes corporate sustainability practices in Japan compared with the international sample of the ICSB 2012 survey. While a number of similarities are observed with regard to the intention, integration and implementation of corporate sustainability, some particularities specific to the Japanese context are also noted. In many instances the Japanese scores are above the international average, which indicates the significance of sustainability among the investigated companies. Most of the external stakeholders are seen to be more promoting for the implementation of corporate sustainability in Japan than in the international sample. This illustrates the harmonious co-existence in society of the Japanese companies with their key stakeholders (Fukukawa and Moon 2004; Vilanova and Tanimoto 2009). Rather than NGOs, which are most strongly promoting corporate sustainability in the international sample, community is found to be most influential in Japan. While companies in Japan have long been associated with the development of communities (Wokutch and Shepard 1999; Fukukawa and Moon 2004), the relatively small role of NGOs in corporate sustainability is also observed by Brucksch and Grünschloß (2008) and Tanimoto (2004). A strong influence of consumers and trade associations in Japan as compared to other countries is also notable.

It can also be seen that companies in Japan prefer different methods and guidelines for their sustainability management. Environmental accounting and reporting, green purchasing, community advisory panel, eco-audit and corporate/employee volunteering, which are frequently used in corporate sustainability in Japan, are less popular in the international sample. Similarly the preference for ISO standards (ISO 14001 and ISO 9000) in Japan is also revealed in the survey. A number of domestic and international factors driving these practices can be identified. The Japanese Government has played a significant role in this respect. For example, environmental accounting and reporting in Japan has been driven by two voluntary government guidelines, namely Environmental Accounting Guidelines and the Environmental Management Accounting Workbook (Kokubu et al. 2003; Kokubu and Nashioka 2005). The green purchasing movement has been enhanced by the Law on Promoting Green Purchasing of 2001. While companies are aware of the positive effects that ISO standards can bring about regarding internal improvement and external legitimacy, especially in the international market, technical and financial support provided by the Japanese Government is also instrumental for the wide acceptance of such standards (Arimura et al. 2005; Nishitani 2009; Welch et al. 2002; Nakamura et al. 2001).

Although external communication (informing stakeholders) is observed as an important measure in sustainability management, public relation and investor relation departments are seen to be relatively less involved in sustainability measurement in Japan compared to the international sample. In addition, the lack of involvement of the personnel/HR department is also notable, given that the companies in Japan have rated employee motivation as the most expected impact of sustainability management. Instead in Japan, some operation-oriented organisational units such as procurement, logistics and quality control are relatively more involved in sustainability measurement. The involvement of procurement and logistics may be due to the strong emphasis of companies on green purchasing and also closely managing transportation issues. It is also striking that Japanese companies along with other social and environmental issues also closely manage biodiversity, which is given less weight in the international sample. The role of the Government and Nippon Keidanren in promoting biodiversity is explained in the Biodiversity Network Japan (2008).

While companies in the international sample highlight internal deficiencies such as lack of personnel and financial capacities as the most important barriers to corporate sustainability, bureaucracy and methodological complexities of sustainability management are considered to be more important in Japan.

In general, similar to their international counterparts, companies in Japan are also highly satisfied with the impact of sustainability management on their competitiveness.

Acknowledgement This research is supported by the Environmental Research and Technology Development Fund (1E-1106) by the Ministry of the Environment, Japan and SPS KAKENHI Grant Number 25285138 and 24830088.

References

Arimura T, Hibiki A, Katayama H (2005) Is a voluntary approach an effective environmental policy instrument? A case for environmental management systems. Resources for the future discussion paper, no 07–31. Resources for the Future, Washington

Brucksch S, Grünschloß C (2008) From environmental accountability to corporate social responsibility? Reflections on the CSR boom in Japan from the perspective of business management and civil society groups. In: Elis V, Lützeler R (eds) Japanstudien 20. Regionalentwicklung und regionale Disparitäten. Iudicium, München, pp 307–329

Burritt R, Schaltegger S, Kokubu K, Wagner M (2003) Environmental management accounting for staff appraisal: evidence from Australia, Germany and Japan. In: Bennett M, Rikhardsson PM, Schaltegger S (eds) Environmental management accounting: purpose and progress. Kluwer, Dordrecht, pp 151–188

Choi S, Aguilera RV (2009) CSR dynamics in south Korea and Japan: a comparative analysis. In: Mallin CA (ed) Corporate social responsibility. A case study approach. Edward Elgar, Cheltenham, pp 123–147

Epstein MJ (2008) Making sustainability work: best practices in managing and measuring corporate social, environmental, and economic impacts. Berrett-Koehler, San Francisco

Fukukawa K (2010) Corporate social responsibility in Japan: a local-global perspective. Working paper, no 10/03. Bradford University School of Management, Bradford

Fukukawa K, Moon J (2004) A Japanese model of corporate social responsibility: a study of website reporting. J Corp Citizenship 14:45–59

Fukukawa K, Teramoto Y (2009) Understanding Japanese CSR: the reflection of managers in the field of global operations. J Bus Ethics 85(1):133–146

Hofstede G (2001) Culture's consequences: comparing values, behaviors, institutions and organizations across nations. Sage Publications, Thousand Oaks

Japan BN (2008) Conserving nature: a Japanese perspective. Biodiversity Network Japan, Tokyo

Kawamura M (2003) Japanese companies launch a new era of CSR management in 2003. NLI Research, Tokyo

Kawamura M (2005a) The evolution of corporate social responsibility in Japan (part 1): parallels with the history of corporate reform. NLI Research, Tokyo

Kawamura M (2005b) The evolution of corporate social responsibility in Japan (part 2): how CSR "swells" have impacted corporate values. NLI Research, Tokyo

Keidanren (2003) Corporate community relations white book. Nippon Keidanren, Tokyo

Keidanren (2010) Charter of corporate behaviour & its implementation guidance (Kigyo-Kodo-Kensho). Keidanren (Japan Business Federation), Tokyo

Keizai D (2004) Corporate social responsibility in Japan: current status and future challenges. CSR survey 2003. Keizai Doyukai (Japan Association of Corporate Executives), Tokyo

Kokubu K, Nashioka E (2005) Environmental management accounting practices in Japan. In: Rikhardsson PM, Bennett M, Bouma JJ, Schaltegger S (eds) Implementing environmental management accounting: status and challenges. Springer, Dordrecht, pp 321–342

Kokubu K, Nashioka E, Saio K, Imai S (2003) Two governmental initiatives on environmental management accounting and corporate practices in Japan. In: Bennett M, Rikhardsson PM, Schaltegger S (eds) Environmental management accounting: purpose and progress. Kluwer, Dordrecht, pp 89–114

Lewin AY, Sakano T, Stephens CU, Victor B (1995) Corporate citizenship in Japan: survey results from Japanese firms. J Bus Ethics 14(2):83–101

METI (Ministry of Economy Trade & Industry) (2004) The new value creation economy and evolving modalities of competition. Ministry of Economy, Trade and Industry, Tokyo

Ministry of the Environment of Japan (2008) Formulation of the third national biodiversity strategy of Japan. In: Biodiversity Network Japan (ed) Conserving nature: a Japanese perspective. Biodiversity Network, Tokyo, pp 60–65

Nakamura M, Takahashi T, Vertinsky I (2001) Why Japanese firms choose to certify: a study of managerial responses to environmental issues. J Environ Econ Manag 42(1):23–52

Nippon Keidanren Committee on Nature Conservation (2008) Initiatives on biodiversity by Nippon Keidanren Committee on nature conservation and its member companies. In: Biodiversity Network Japan (ed) Conserving nature: a Japanese perspective. Biodiversity Network, Tokyo, pp 56–59

Nishitani K (2009) An empirical study of the initial adoption of ISO 14001 in Japanese manufacturing firms. Ecol Econ 68(3):669–679

Schaltegger S, Beständig U (2010) Corporate biodiversity management handbook. BMU/GTZ/CSM, Berlin/Eschborn/Lüneburg. http://www.business-and-biodiversity.de/en/activities/biodiversity-management/handbook. Accessed 22 Oct 2013

Schaltegger S, Burritt R (2005) Corporate sustainability. In: Folmer H, Tietenberg T (eds) The international yearbook of environmental and resource economics 2005/2006. Edward Elgar, Cheltenham, pp 185–222

Schaltegger S, Lüdeke-Freund F, Hansen E (2012) Business cases for sustainability. The role of business model innovation for corporate Sustainability. Int J Innov Sustainable Dev 6(2):95–119

Schaltegger S, Harms D, Hörisch J, Windolph SE, Burritt R, Carter A, Truran S, Crutzen N, Ben Rhouma A, Csutora M, Tabi A, Kokubu K, Kitada H, Haider MB, Kim J-D, Lee K-H, Moneva JM, Ortas E, Álvarez-Etxeberria I, Daub C-H, Schmidt J, Herzig C, Morelli J (2013) International corporate sustainability barometer: a comparative study of 11 countries. Centre for Sustainability Management, Lüneburg

Tanimoto K (2004) CSR keiei: Kigyō no shakai sekinin to suteikuhorudā [CSR management: the social responsibility and stakeholders of companies]. Chūō Keizaisha, Tokyo

Tanimoto K, Suzuki K (2005) Corporate social responsibility in Japan: analyzing the participating companies in the Global Reporting Initiative. Working paper, no 208. Stockholm School of Economics, Stockholm

Vilanova M, Tanimoto K (2009) Best environmental practices of AEON: a case study. Greening Retail, Canada

Welch EW, Mori Y, Aoyagi-Usui M (2002) Voluntary adoption of ISO 14001 in Japan: mechanisms, stages and effects. Bus Strategy Environ 11(1):43–62

Wokutch RE (1990) Corporate social responsibility Japanese style. Executive 4(2):56–74

Wokutch RE, Shepard JM (1999) The maturing of the Japanese economy: corporate social responsibility implications. Bus Ethics Q 9(3):527–545

Zadek S (2007) The civil corporation. Earthscan, London/Sterling

Chapter 10
The Current Status of Korean Corporate Sustainability Management

Jong-Dae Kim and Ki-Hoon Lee

Abstract This chapter provides key findings and particularities of the Korean Corporate Sustainability Barometer survey. The majority of companies in the Korean sample are large-sized and belong to the manufacturing sector. As external drivers, NGOs and the media/public are the stakeholders most strongly impacting corporate sustainability implementation. Furthermore, top management commitment and legislation are found to be important factors promoting sustainability implementation. Notably, Korean companies indicate that the local community is also a strong driver. Korean companies consider legislation and lack of financial capacities to be major barriers to the implementation of corporate sustainability management. With regard to the integration of sustainability into the core business and organisation, the majority of Korean companies show that they connect sustainability to most or all segments of their core business. This indicates a high level of sustainability integration into the core business and organisation of Korean companies. Driven by formal and informal institutional forces, Korean companies are aware of the importance of sustainability, and they actively adopt sustainability management tools and approaches to implement corporate sustainability management.

J.-D. Kim
College of Business Administration, Inha University, Incheon, Republic of Korea
e-mail: jdk@inha.ac.kr

K.-H. Lee (✉)
Griffith Business School, Griffith University, Brisbane, Australia
e-mail: ki-hoon.lee@griffith.edu.au

S. Schaltegger et al. (eds.), *Corporate Sustainability in International Comparison*,
Eco-Efficiency in Industry and Science 31, DOI 10.1007/978-3-319-06227-3_10,
© Springer International Publishing Switzerland 2014

10.1 Introduction

10.1.1 The Korean Context

South Korea, officially the Republic of Korea, is a relatively small country in the southern part of the Korean Peninsula. The land mass of the country is 100,210 km^2 and it has a population of approximately 48.9 million. The official language is Korean and the currency is the Korean won. In regard to labour and educational institutions, Korea is characterised by a skilled labour market and, as a result, a large supply of human capital (OECD 2012). Due to its fast-growing economy among OECD economies, Korea has had one of the highest growth levels of greenhouse gas emissions in the OECD countries since 1990. In terms of political institutions, laws and regulations promoting environmental, social and economic sustainability are more prevalent in Korea, a characteristic that limits the ability of companies to gain competitive advantages through corporate sustainability activities and to generate high sustainability performance (Lee 2013). Major Korean industries, including the construction, steel, automobile, electronics, semiconductor, shipbuilding and heavy manufacturing industries, are energy intensive. Nowadays, Korean business organisations adopt and implement a range of sustainability management activities (FKI 2011).

The 2007 IPCC report delivered the urgent message "more with less" in terms of climate change and national and international economic productivity. The current major sustainability management challenge at the national level is to reduce carbon emissions and create new opportunities for industries to boost a green economy. With a vision of "low carbon, green growth", Korea established a Green Growth Strategy in 2009 aiming at 30 % emission reductions by 2020, implying a 4 % emission cut from the 2005 level (SGERC 2012; Lee 2013). The strategy also includes a Five-Year Plan (2009–2013) containing about 600 projects and a total budget of 108.7 trillion won (10 % of 2009 GDP) to promote green growth. Public research and development (R&D) accounts for 11 % of the total GDP, motivated by the need to overcome market failures related to the high degree of uncertainty and long time horizons in green innovation hindering private-sector research (Green Growth Committee 2009). In 2009, private firms were involved in nearly two-thirds of the 4,732 R&D projects in the Five-Year Plan. However, a much greater business involvement in terms of financial contribution is needed to promote and advance green research and make it a key driver of innovation in the private industry sector (OECD 2012). In the 1990s and 2000s, multinational Korean enterprises such as Samsung, LG, Hyundai, Kia, Posco, Daewoo and Hynix experienced environmental and social pressures from the markets in which they operated. Since the Korean economy underwent financial reforms in 1997, the government has placed much higher environmental, economic and social compliance standards on Korean industries in domestic markets than those accepted internationally. As Lee and Kim (2014) point out, government legislation and policy are one of the major factors motivating firms to adopt and implement sustainability and corporate social responsibility activities in Korea. Thus, it becomes apparent that Korean companies face sustainability challenges and that many managers agree that sustainability can offer new business opportunities, but these are not easy to implement.

At an industry and firm level, growing attention has been paid to corporate sustainability by the business world. A global survey by McKinsey (2008) found that climate change and carbon management are clearly a reality of business today. Similarly, Hoffman and Woody (2008) noted that a climate strategy is a business strategy for leading global firms. The UN Global Compact and Accenture CEO study 2010 pointed out that 93 % of the 766 participating chief executive officers (CEOs) from over 100 countries and 25 industries worldwide agreed that "sustainability is more important than ever to the future success of their businesses" (Lacy et al. 2010:16). Similarly, Lee (2012a) and Schaltegger et al. (2012) highlight the roles of drivers and business models to create business cases for sustainability. For example, Schaltegger et al. (2012) provide an integrated framework of sustainability strategy, business case drivers and business model innovation in order to overcome organisational barriers or hurdles which many firms face during the identification and implementation of sustainability management. More recently, Eccles and Serafeim (2013) conducted an environmental, social and governance (ESG) performance study to explore sustainability strategy by surveying about 3,000 companies internationally. Key findings from their study show that firms need to focus strategically on "the most important ESG issues which have the greatest impact on the firm's ability to create shareholder value; and to produce major innovations in products, processes, and business models that prioritize those concerns" (Eccles and Serafeim 2013:52). Thus, recent studies and surveys postulate that firms need to adopt and implement strategically-focused sustainability in order to avoid negative impacts on shareholder value and firm performance (Porter and Kramer 2006, 2011; Lee 2012b; Cheng et al. 2014). In order to investigate this important issue, our study attempts to answer the following key questions. Why do companies manage sustainability (intention)? To what extent do companies embed sustainability in their core business and in their organisation (integration)? And how is corporate sustainability operationalised (implementation)?

10.1.2 The Korean Sample

The Korean survey is part of the International Corporate Sustainability Barometer coordinated by the Centre for Sustainability Management (CSM) in Lüneburg, Germany (see Schaltegger et al. 2013). The statistics of participating companies indicate that the Korean sample mostly contains large-sized companies according to the number of employees (about 60 % of companies have between 1,001 and 10,000 employees; 37.5 % of companies have between 10,001 and 100,000 employees). As Fig. 10.1 shows, this sample size is similar to the international survey sample (which

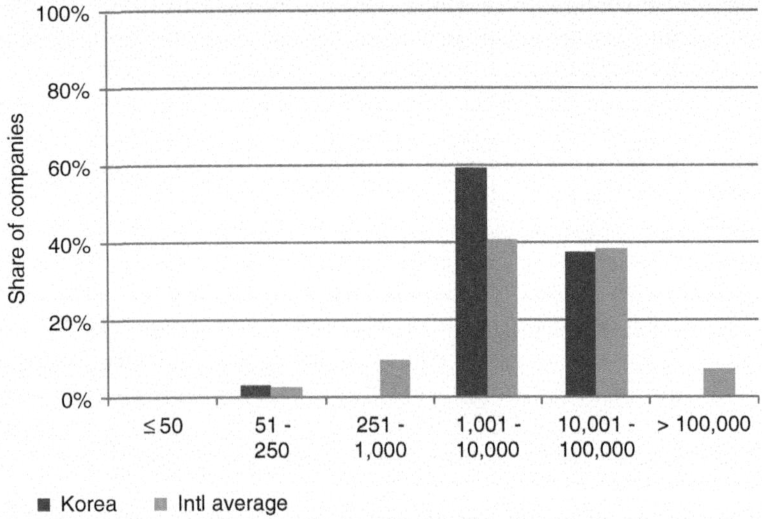

Fig. 10.1 Number of employees ($n_{KOR} = 32$, $n_{INT} = 465$)

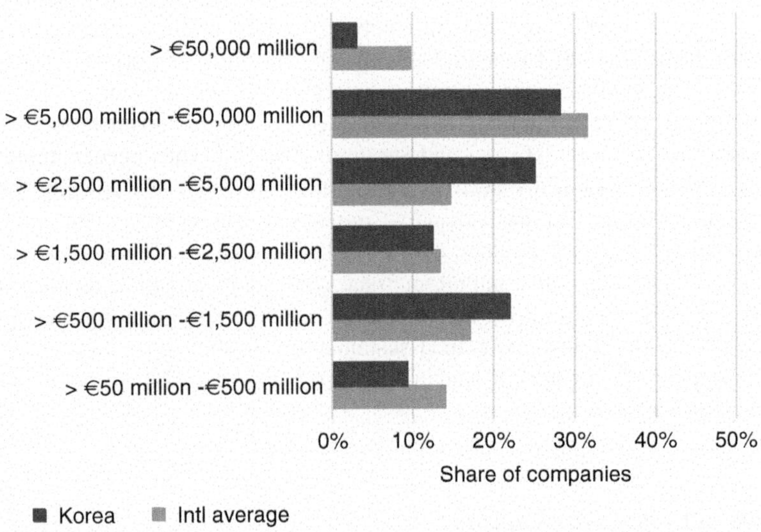

Fig. 10.2 Annual revenue ($n_{KOR} = 32$, $n_{INT} = 468$)

shows that about 41 % of companies have between 1,001 and 10,000 employees; 38.7 % companies have between 10,001 and 100,000 employees).

In addition, revenue information in Fig. 10.2 shows that the majority of participating Korean companies are categorised as earning between more than €5,000 and €50,000 million (28.1 %) and between more than €2,500 and €5,000 million (25.0 %), per annum respectively. This revenue information is also similar to the international average of between more than €5,000 and €50,000 million (31.4 %) and between more than €2,500 and €5,000 million (14.5 %), per annum respectively.

In this study, the core economic sectors of the sample companies are (i) industry, capital goods & building (53.1 %); (ii) consumer goods, trade & logistics (21.9 %); (iii) finance & services (12.5 %); (iv) commodities, auxiliary material, energy, chemical & pharmaceutical industry (12.5 %). In the international sample finance & services amount to 32.1 %; commodities, auxiliary material, energy, chemical & pharmaceutical industry 23.7 %; industry, capital goods & building 22.6 %; and consumer goods, trade & logistics 21.6 %. Since Korea has a strong manufacturing and research and development sector, there are some notable differences in terms of core business.

10.2 Analysis

10.2.1 Intention

10.2.1.1 Motivation (External)

Not surprisingly, similar tendencies between the Korean and the international average data are observed with regard to the impacts of external stakeholders (see Fig. 10.3). There are however some notable differences between the Korean and the international sample data. Most companies in both the Korean and the international samples agree that NGOs/environmental/social organisations (Korea 3.93; international 4.03), and media/public (Korea 3.84; international 3.84) are the external stakeholders which most strongly impact the implementation of sustainability in a company. However, there are some minor differences in the way in which international authorities and the community impact the implementation of corporate sustainability. The international average indicates that international authorities (e.g. United Nations) are also very strong drivers of the implementation of corporate sustainability (Korea 3.55; international 3.70), whereas Korean companies indicate that community is a stronger driver for the implementation of corporate sustainability (Korea 3.73; 3.67). Furthermore, banks/creditors (Korea 2.94; international 3.11) and insurance companies (Korea 2.81; international 3.13) are considered 'passive' or neutral with regard to the implementation of corporate sustainability in both the international average and the Korean samples. This may indicate the fact that banks and insurance companies do not play a major role in corporate sustainability implementation, neither in Korea nor on international average.

As shown in Fig. 10.4, this survey revealed that commitment by top management (Korea 4.22; International 4.28) and legislation (Korea 3.97; International 3.95) play a major role in promoting corporate sustainability implementation.

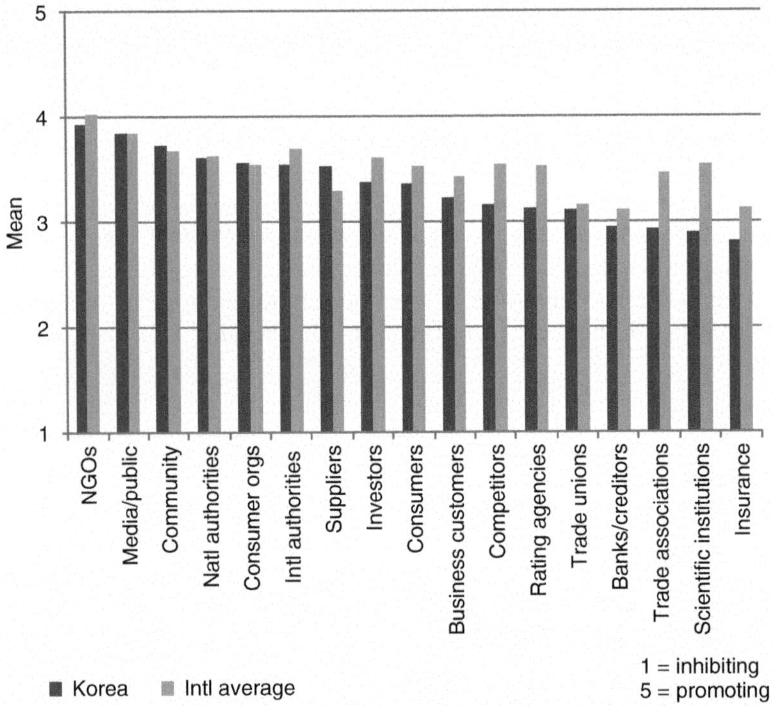

Fig. 10.3 Impact of external stakeholders (n_{KOR} ranging from 28 to 32, n_{INT} ranging from 393 to 450)

In summary, top management commitment, NGOs, media and legislation are the main drivers for the implementation of corporate sustainability in Korea as well as on international average.

10.2.1.2 Barriers (Internal and External)

In identifying barriers (inhibiting factors) to the implementation of corporate sustainability, interesting differences between the Korean and the international samples can be found (see Fig. 10.5). The Korean sample considered legislation (3.26) and lack of financial capacities (3.13) as major barriers to the implementation of corporate sustainability, while the international average evaluated lack of personnel capacities (3.11) and lack of financial capacities (3.09) as being the most important barriers to the implementation of corporate sustainability. In both the Korean and the international sample, lack of financial capacities is identified as an important barrier for corporate sustainability. Interestingly, the Korean sample indicates that legislation is the most important barrier, while the international average (2.91) regards legislation as a non-major issue. This may indicate that legal or legislative

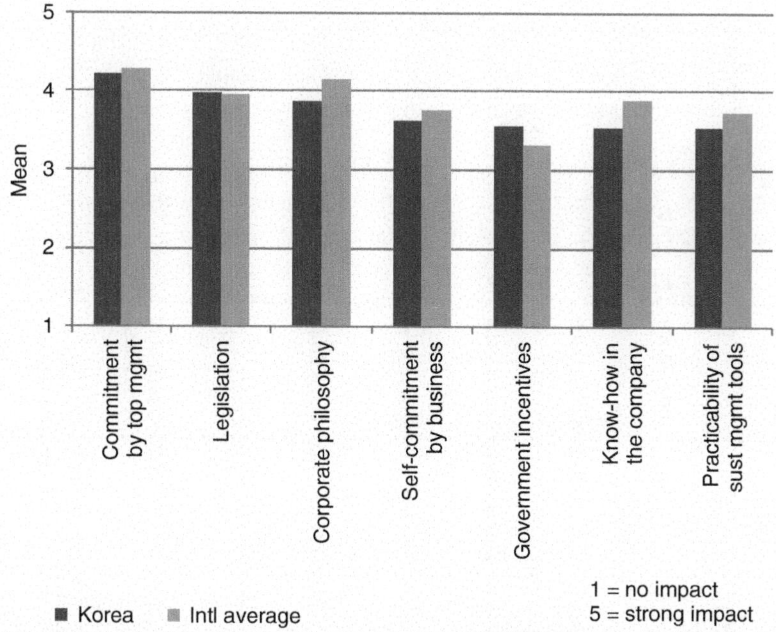

Fig. 10.4 Promoting factors ($n_{KOR} = 32$, n_{INT} ranging from 446 to 460)

institutional factors play an important role in promoting or inhibiting corporate sustainability in Korea compared to the international average.

For the implementation of corporate sustainability, financial capacities, government incentives, personnel capacities and methodological issues of sustainability management are very important factors to consider. Besides legislation and lack of financial capacities, the participating Korean companies consider it difficult to implement corporate sustainability due to methodological complexity. It is probable that, in the case of Korea, there are too many different methods and tools to implement corporate sustainability, and new tools and methods to improve the implementation of corporate sustainability are introduced relatively quickly. This methodological complexity issue can be caused by government legislation or international standards, and higher institutional pressure may be put on Korean firms to adopt and implement the many tools and methods necessary for corporate sustainability.

10.2.1.3 Issues

Environmental Issues

As can be seen in Fig. 10.6, the Korean and the international samples agree that energy consumption (Korea 4.69; international 4.57) and emissions/wastewater/

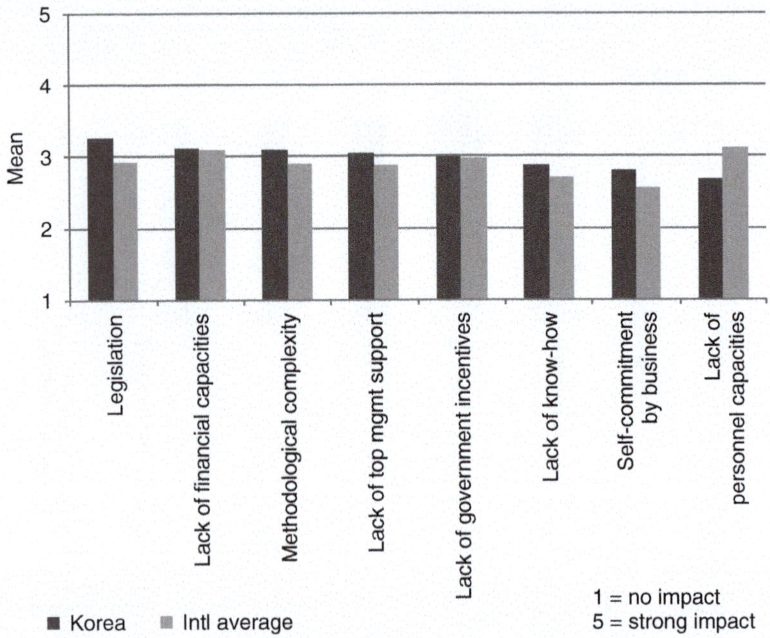

Fig. 10.5 Inhibiting factors (n_{KOR} ranging from 31 to 32, n_{INT} ranging from 438 to 449)

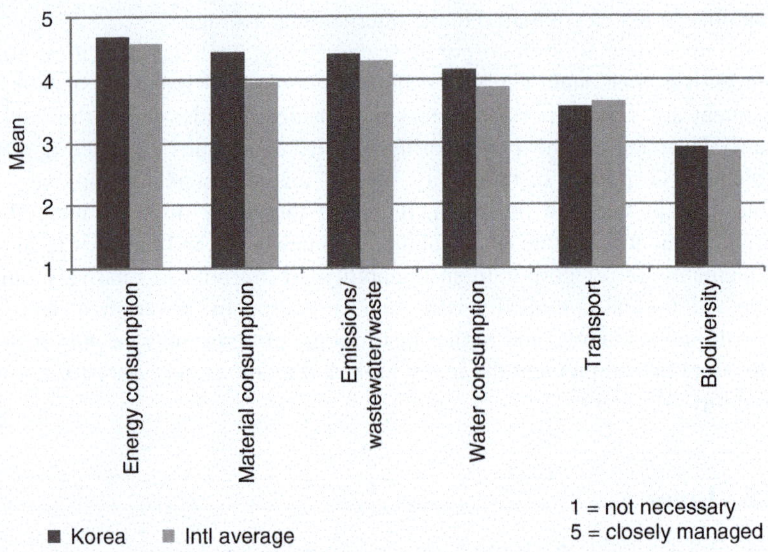

Fig. 10.6 Environmental issues (n_{KOR} = 32, n_{INT} = 455-463)

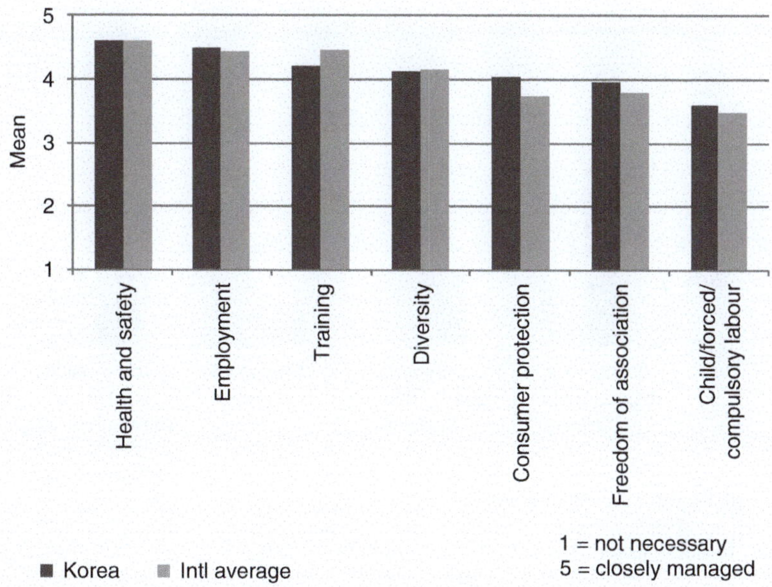

Fig. 10.7 Social issues (n_{KOR} ranging from 30 to 31, n_{INT} ranging from 442 to 461)

waste (Korea 4.41; international 4.29) are major sustainability issues needing management. Similar tendencies can also be found for water consumption and material consumption.

Social Issues

Most companies in the survey agree that workplace/employment (Korea 4.48; international 4.44), occupational health and safety (Korea 4.61; international 4.61) as well as training and development (Korea 4.23; international 4.47) are the key issues being managed (see Fig. 10.7). Furthermore, the international average and the Korean sample state that they closely manage diversity and equal opportunity (Korea 4.13; international 4.16). Consumer protection (Korea 4.03; international 3.73), however, is more closely managed in Korea.

10.2.2 Integration

10.2.2.1 Connection to Core Business

As shown in Fig. 10.8, the Korean sample has a focus on the manufacturing industry sector including industry, capital goods & building, while the international average

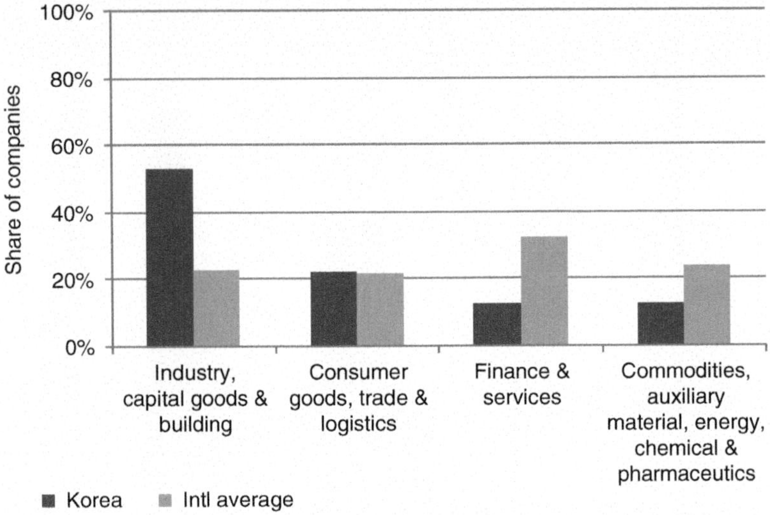

Fig. 10.8 Business sectors ($n_{KOR} = 32$, $n_{INT} = 468$)

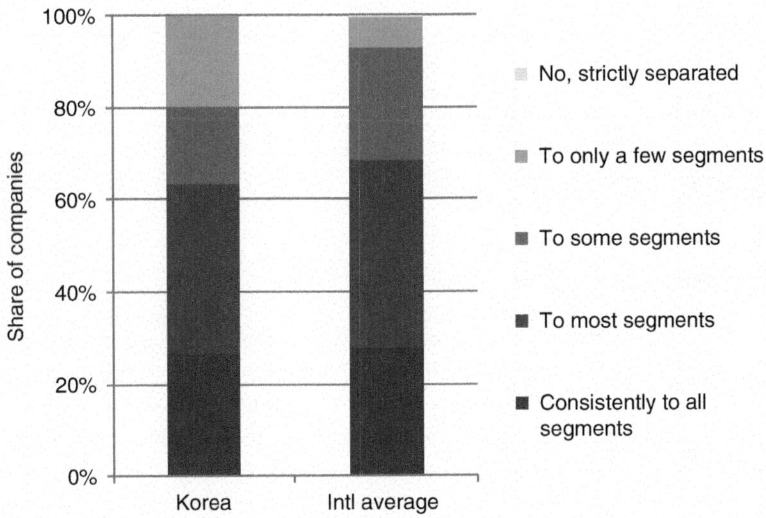

Fig. 10.9 Connection to the core business ($n_{KOR} = 30$, $n_{INT} = 457$)

has a relatively high percentage of finance & service sector and commodities, energy, chemical & pharmaceutical industry sectors.

In order to achieve sustainable organisation, it is important to link environmental and social improvements to economic success and to integrate the engagement for sustainability into the core business. Figure 10.9 demonstrates the level of connections to the core business in the Korean and international samples. As Fig. 10.9 shows, most companies in the survey agree that they connect sustainability to

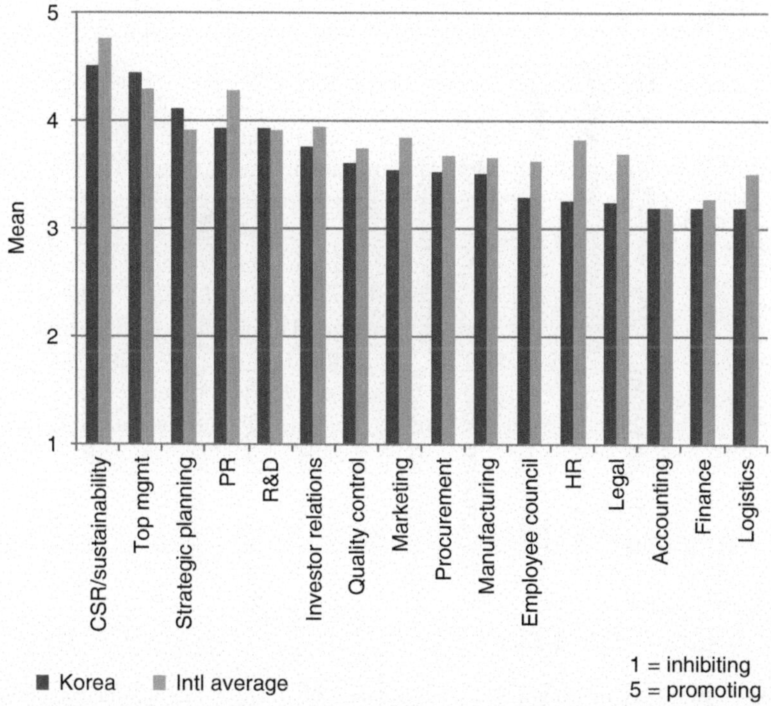

Fig. 10.10 Impact of internal stakeholders/organisational units (n_{KOR} ranging from 27 to 32, n_{INT} ranging from 325 to 460)

most segments of their core business (Korea 36.7 %; international 40.7 %). Importantly, both survey groups also show that about one-quarter of the companies consistently connect sustainability to *all* segments of their core business (Korea 26.7 %; international 27.8 %). It is important to note that a majority of Korean and international sample companies integrate sustainability into their core business to a certain extent.

10.2.2.2 Involvement of Organisational Units

As Fig. 10.10 shows, most companies in the Korean and the international samples agree that top management (Korea 4.44; international 4.29), and CSR/sustainability (including environment, health and safety; Korea 4.52; international 4.76) are the strongest internal drivers of sustainability implementation in a company. However, there are some differences in public relations (PR)/corporate communication and strategic planning.

The international average indicates that PR is also a very strong driver of corporate sustainability implementation (Korea 3.94; international 4.28). Notably, in the Korean sample strategic planning receives a higher value than PR, while the

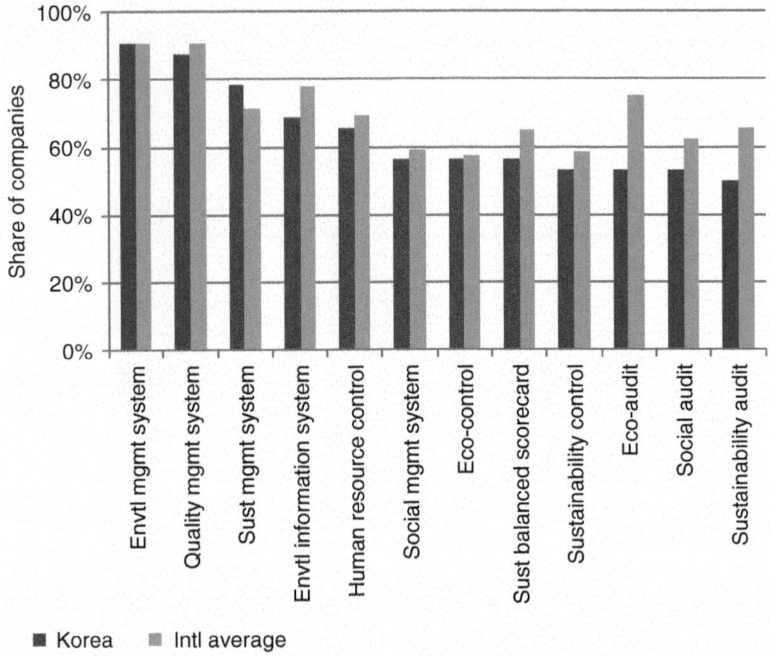

Fig. 10.11 Awareness of control and management tools ($n_{KOR} = 32$, n_{INT} ranging from 467 to 468)

international average value for PR is higher than the value for strategic planning. In part, strategic planning in many Korean companies plays a ‘control tower’ role in allocating resources including budget, labour, etc. In this survey, Korean companies indicate that strategic planning is a stronger driver of corporate sustainability implementation (Korea 4.10; international 3.91). In summary, Korean companies have three top motivating factors, and these are top management, CSR/sustainability and strategic planning, while the international sample companies have top management, CSR/sustainability, and public relations/corporate communication as the top three drivers of corporate sustainability.

10.2.3 Implementation

10.2.3.1 Sustainability Management Tools (Knowledge)

Control and Management

As shown in Fig. 10.11, environmental and quality tools are very popular in the Korean and the international samples. Most companies in the survey agree that environmental management systems (Korea 90.6 %; international 90.8 %) and

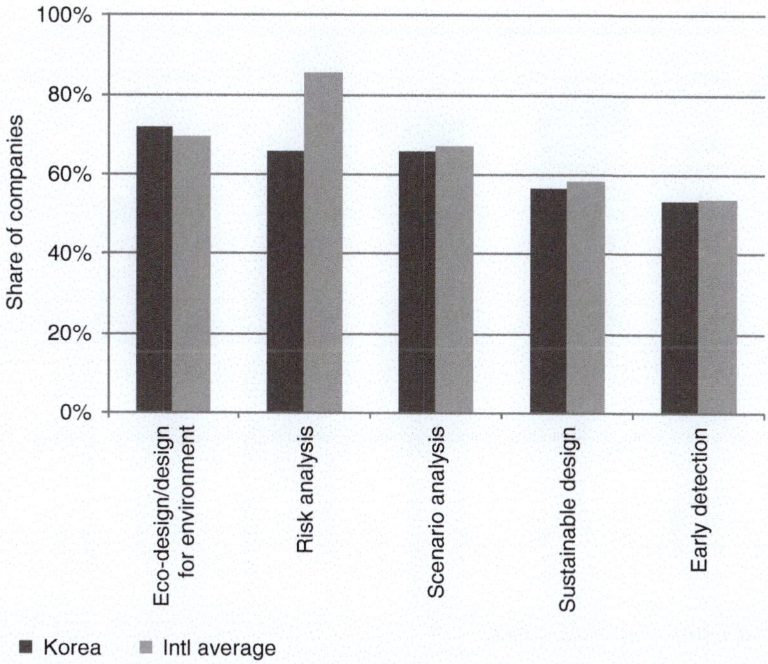

Fig. 10.12 Awareness of development and planning tools ($n_{KOR} = 32$, $n_{INT} = 467$)

quality management systems (QMS) (Korea 87.5 %; international 90.8 %) are two well-known tools in the control and management category of sustainability management tools. In addition, sustainability management systems are popular both in the Korean and the international samples (Korea 78.1 %; international 71.3 %). Eco-audits, however, are better known on international average than in Korean companies (Korea 53.1 %; international 75.0 %).

Development and Planning

In the category of development and planning, eco-design/design for the environment (Korea 71.9 %; international 69.6 %), risk analysis (Korea 65.6 %; international 85.4 %), and scenario analysis (Korea 65.6 %; international 67.2 %) are well-known in both the Korean and the international samples (see Fig. 10.12). In particular, the international sample indicates that risk analysis (85.4 %) is a very well-known sustainability tool.

Purchasing and Producing

In the purchasing and producing category, green purchasing (Korea 75.0 %; international 77.6 %), green supply chain management (Korea 68.8 %; international

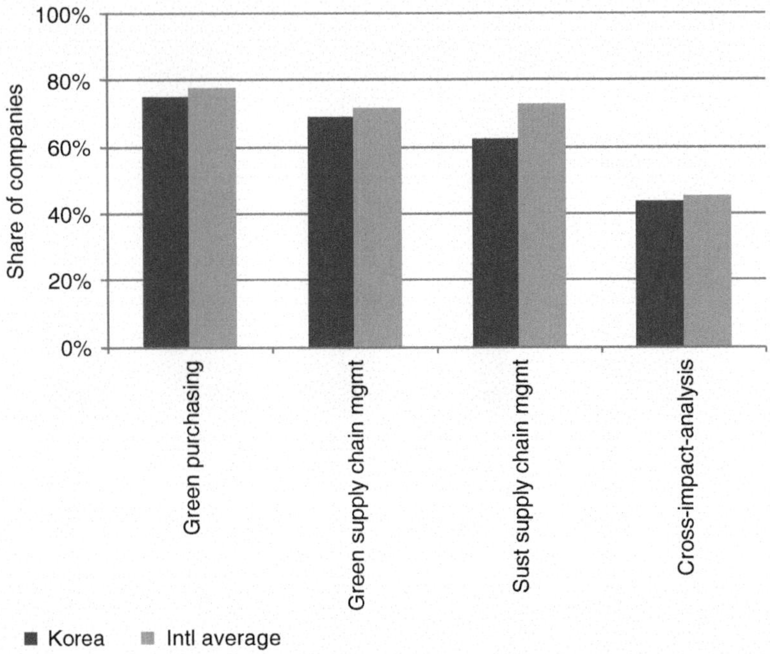

Fig. 10.13 Awareness of purchasing and producing tools ($n_{KOR} = 32$, n_{INT} ranging from 467 to 468)

71.8 %), and sustainable supply chain management (Korea 62.5 %; international 72.6 %) are popular in both the Korean and the international samples. No notable differences are observed in this category (see Fig. 10.13).

Measuring and Comparing

As Fig. 10.14 shows, environmental tools are popular both in the Korean and the international samples. In particular, material flow analysis (Korea 65.6 %; international 62.1 %), eco-efficiency analysis (Korea 65.6 %; international 57.5 %), eco-balance/life cycle assessment (Korea 65.6 %; international 71.8 %), environmental cost accounting (Korea 65.6 %; international 58.8 %), environmental accounting (Korea 53.1 %; international 51.6 %) and sustainability indicators (Korea 56.3 %; international 63.8 %) are well-known tools in both the Korean and the international samples. Eco-indicators (Korea 50.0 %; international 71.2 %) and product carbon footprint (Korea 62.5 %; international 77.1 %) are relatively more popular in the international sample.

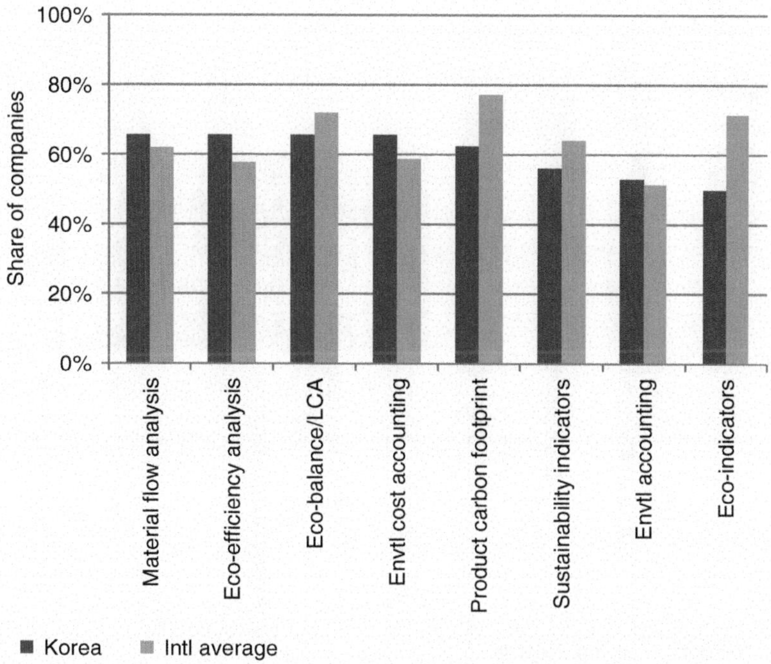

Fig. 10.14 Awareness of measuring and comparing tools ($n_{KOR} = 32$, n_{INT} ranging from 467 to 468)

Communication and Marketing

Figure 10.15 shows that the Korean and the international samples are similarly aware of corporate donation and giving (Korea 68.8 %; international 74.9 %), environmental reporting (Korea 78.1 %; international 84.6 %) and sustainability reporting (Korea 81.3 %; international 86.1 %). However, there are some notable differences in environmental mission statements (Korea 50.0 %; international 78.4 %) and sustainability mission statements (Korea 46.9 %; international 71.7 %). The relatively high level of use of communication and marketing tools in the international sample is consistent with the findings of a high level of public relations and communication involvement in the motivation and driving factors within the international sample, as shown above.

Employee Motivation and Involvement

As Fig. 10.16 shows, most companies in the Korean and the international samples know the proposal system (Korea 81.3 %; international 74.7 %) and corporate/employee volunteering (Korea 81.3 %; international 79.7 %). However, there are some notable differences in flexible working time (Korea 71.9 %; international

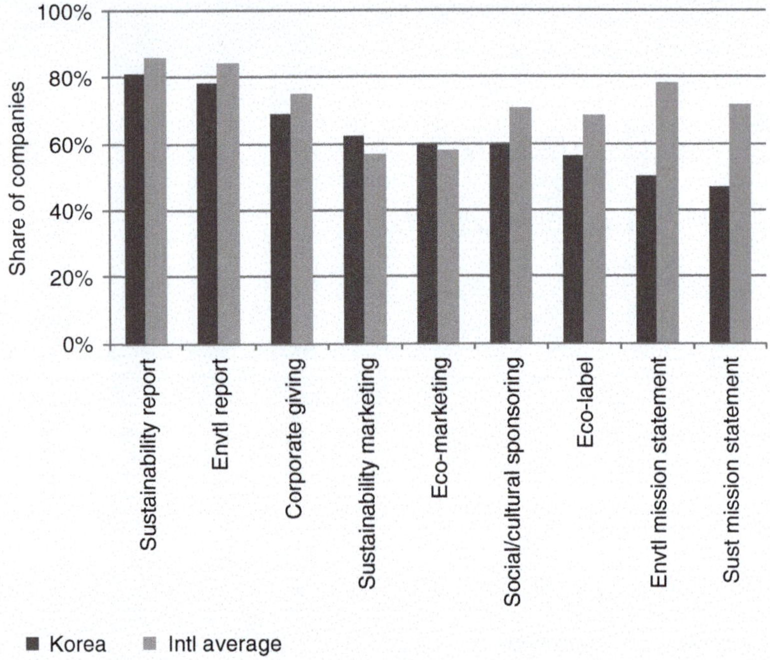

Fig. 10.15 Awareness of communication and marketing tools ($n_{KOR} = 32$, n_{INT} ranging from 467 to 468)

91.2 %), incentive systems (Korea 65.6 %; international 85.0 %) and further education (Korea 62.5 %; international 84.8 %). These last three tools are more popular on international average than in the Korean sample.

Stakeholder-Oriented and Other Methods

Most companies in the Korean and the international samples know the tools corporate citizenship (Korea 71.9 %; international 72.4 %), stakeholder dialogue/forums (Korea 68.8 %; international 75.2 %), and emission certificate trading (Korea 65.6 %; international 61.7 %). No notable differences are observed in this category (see Fig. 10.17).

10.2.3.2 Sustainability Management Tools (Application)

Control and Management

There are general similarities in terms of the actual application of sustainability management tools in the category of control and management (see Fig. 10.18). In

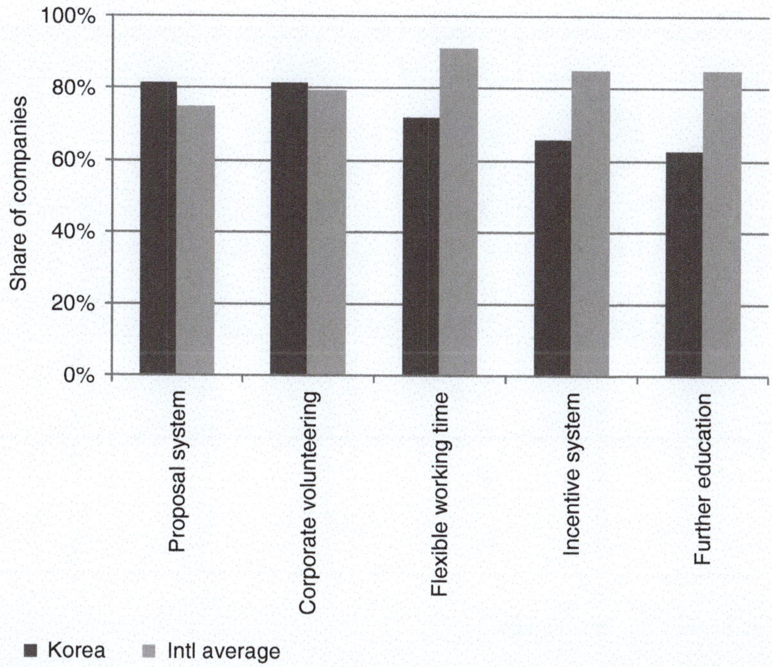

Fig. 10.16 Awareness of employee motivation and involvement tools ($n_{KOR} = 32$, $n_{INT} = 467$)

particular, environmental and quality tools are broadly applied in the Korean and the international samples. Both samples agree that environmental management systems (Korea 84.4 %; international 77.6 %) and quality management systems (QMS) (Korea 78.1 %; international 75.6 %) are the two main tools applied in the control and management category. In comparison with the knowledge of sustainability management tools, a relatively high level of application of control and management tools can be observed in Korea.

Development and Planning

Figure 10.19 demonstrates that eco-design/design for the environment (Korea 56.3 %; international 39.0 %) and risk analysis (Korea 43.8 %; international 70.4 %) are relatively broadly applied sustainability management tools in the Korean and the international samples. Interestingly, it can also be noted that the actual application level in development and planning is relatively low compared to the knowledge of these tools. This indicates that there are some gaps between knowledge and implementation of sustainability management tools.

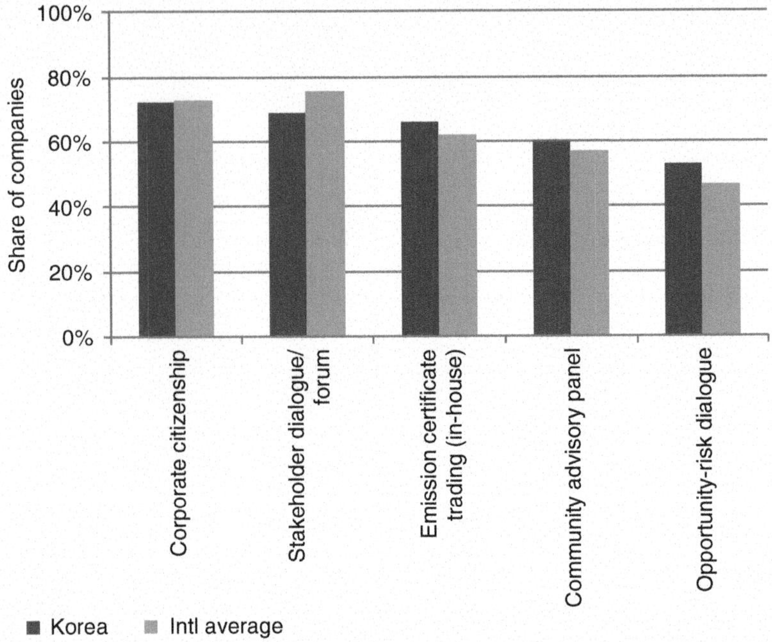

Fig. 10.17 Awareness of stakeholder-oriented and other tools ($n_{KOR} = 32$, n_{INT} ranging from 467 to 468)

Purchasing and Producing

As shown in Fig. 10.20, green purchasing (Korea 59.4 %; international 40.2 %) and sustainable supply chain management (Korea 46.9 %; international 33.8 %) are relatively popular sustainability management tools in both the Korean and the international samples. There is a notable difference for green supply chain management (Korea 53.1 %; international 26.5 %), which is an interesting finding because there are high levels of knowledge of green supply chain management, but its actual application is very low on international average.

Measuring and Comparing

Environmental tools are in general applied by less than half of the companies in the Korean and the international samples (see Fig. 10.21). Eco-efficiency analysis (Korea 40.6 %; international 25.6 %), eco-balance/life cycle assessment (Korea 56.3 %; international 38.2 %), product carbon footprint (Korea 50.0 %; international 37.2 %), and environmental cost accounting (Korea 31.3 %; international 23.1 %) are applied to very different degrees by the Korean companies and on international average. Interestingly, the use of eco-indicators is relatively popular in the international sample.

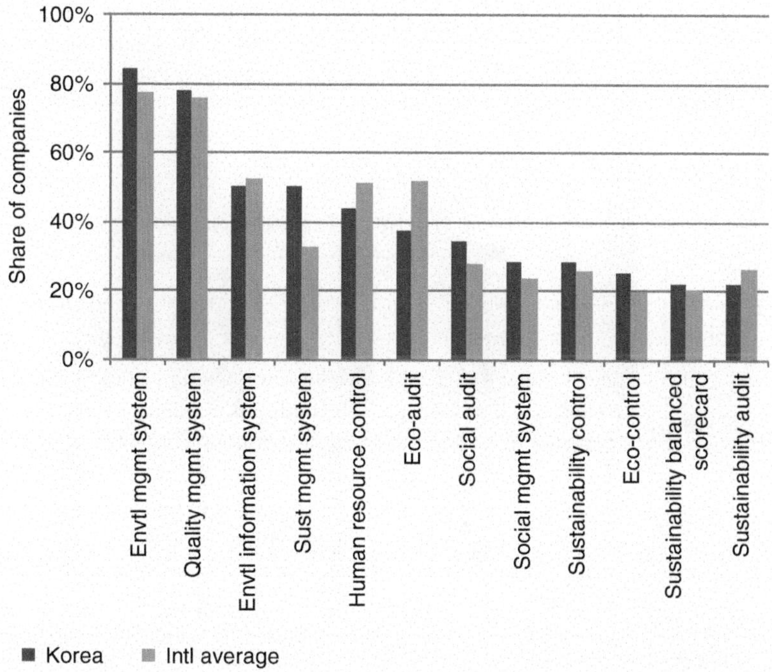

Fig. 10.18 Application of control and management tools ($n_{KOR} = 32$, n_{INT} ranging from 467 to 468)

Communication and Marketing

In the communication and marketing category, most companies in the Korean and the international samples apply corporate giving (Korea 59.4 %; international 61.2 %), environmental reporting (Korea 56.3 %; international 57.3 %), and sustainability reporting (Korea 62.5 %; international 63.2 %). One notable difference in this category can be observed for environmental mission statement (Korea 37.5 %; international 58.5 %; see Fig. 10.22).

Employee Motivation and Involvement

Figure 10.23 shows that a majority of the companies in the Korean and the international samples apply a proposal system (Korea 68.8; international 55.9 %), corporate/employee volunteering (Korea 68.8 %; international 59.5 %) and incentive systems (Korea 50.0 %; international 67.9 %) in the employee motivation and involvement category. However, there are some notable differences in flexible working times (Korea 50.0 %; international 79.0 %) and further education (Korea 50.0 %; international 76.7 %). It is interesting to note that the actual application level

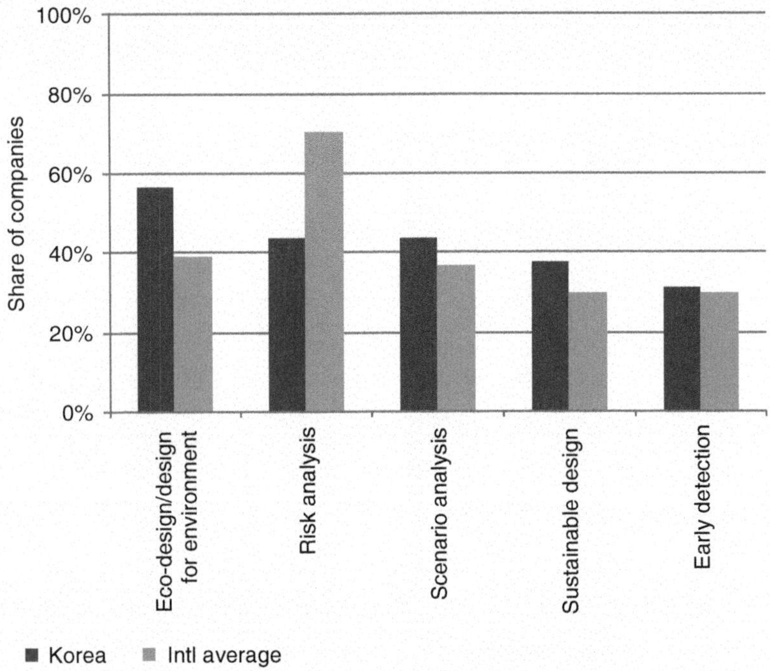

Fig. 10.19 Application of development and planning tools ($n_{KOR} = 32$, $n_{INT} = 467$)

in the employee motivation and involvement category is relatively low compared to knowledge in this category, indicating that there are some gaps between knowledge and implementation of sustainability management tools.

Stakeholder-Oriented and Other Methods

Several companies in the Korean and the international samples state that they apply corporate citizenship (Korea 50.0 %; international 46.7 %) and stakeholder dialogue/forum (Korea 50.0 %; international 45.9 %; see Fig. 10.24). Notably, the application of emission certificate trading (Korea 40.6 %; international 17.1 %) is more widespread in Korea than on international average. It is important to note that again the actual application in the stakeholder-oriented and other methods category is relatively low compared to the knowledge of tools in this category, indicating that there are major gaps between knowledge and implementation.

As shown in Table 10.1, the top three standards or guidelines for both the Korean and the international sample include ISO 14001 (Korea 81.3 %; international 73.1 %), ISO 9000 (Korea 68.8 %; international 66.7 %), and GRI guidelines (Korea 53.1 %; international 52.8 %). Notably, a relatively high share of Korean companies

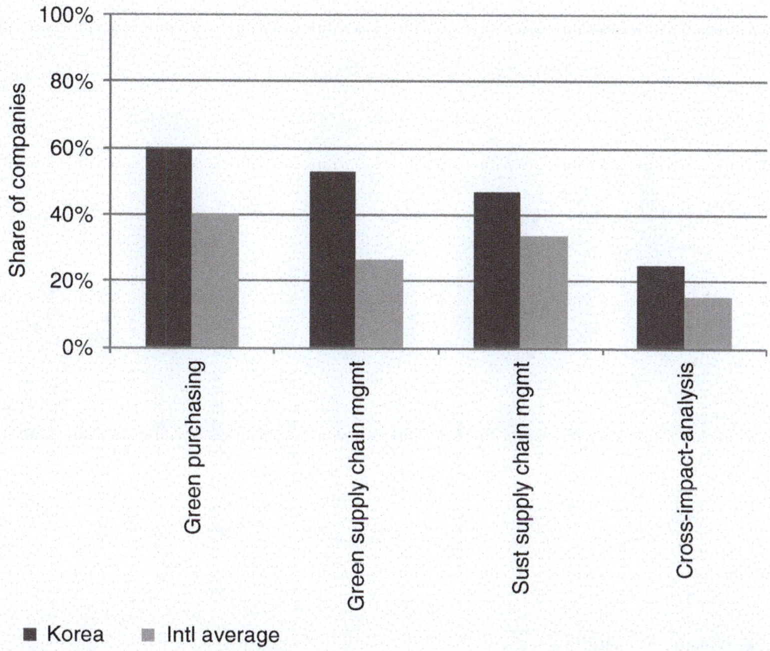

Fig. 10.20 Application of purchasing and producing tools ($n_{KOR} = 32$, n_{INT} ranging from 467 to 468)

also applies OHSAS 18001/BS 8800 (50.0 %), ISO 26000 (43.8 %) and UN Global Compact (43.8 %), while the international sample applies a relatively low level of these standards/guidelines.

10.3 Conclusion

This chapter presents the findings of the International Corporate Sustainability Barometer survey in Korea. Corporate sustainability management is a relatively new phenomenon for Korean companies. Driven by formal and informal institutional forces, many Korean companies demonstrate relatively high levels of commitment to and implementation of corporate sustainability. Government policy and legislation play a key role in promoting and inhibiting corporate sustainability management. Furthermore, motivation and commitment by top management are important for the implementation of sustainability management.

As Porter and van der Linde (1995) argued, properly designed environmental policy and regulations may encourage environmental innovation, promote environmental activities and enhance environmental and economic performance. As this

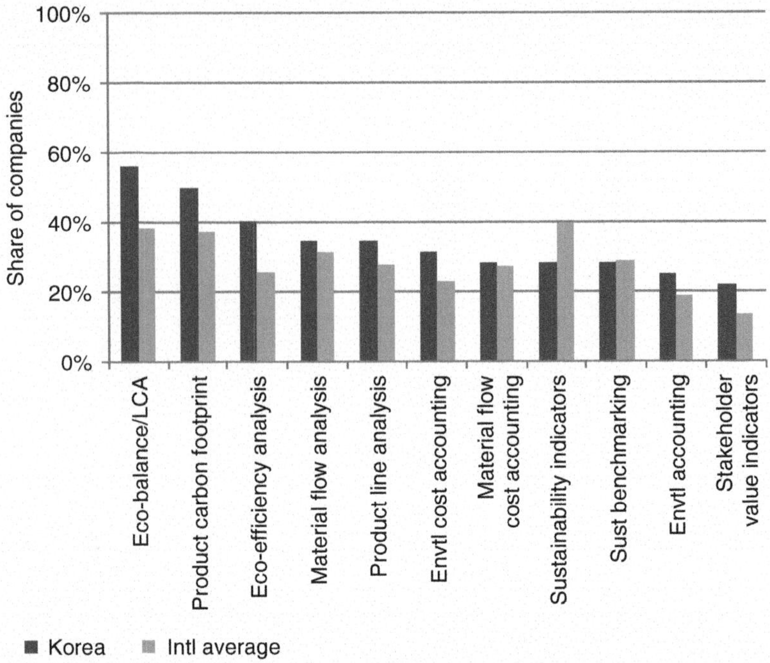

Fig. 10.21 Application of measuring and comparing tools ($n_{KOR} = 32$, n_{INT} ranging from 467 to 468)

chapter shows, Korean government initiatives, legal acts and incentives are important factors in the acceleration of corporate sustainability management in industries. Since Korea adopted the Green Growth Strategy in 2009 and implemented the Five-Year Plan (2009–2013) with public spending of 2 % of GDP per year to promote green growth, Korea has invested a total budget of 108.7 trillion won (10 % of 2009 GDP) for about 600 green innovation projects. Although it is too early to say that Korea's progress on green innovation and green growth is successful, it has become clear that Korean government policy and legislation play a key role in promoting and facilitating corporate sustainability management.

This survey delivers clear data and outcomes about how and to what extent Korean firms integrate corporate sustainability into their core businesses and into their organisations, two suitable indicators of the level of commitment and involvement of business organisations in creating business cases for sustainability. More than 60 % of the Korean companies in this survey study show that they connect sustainability to most or all segments of their core business. This important finding indicates that a high level of sustainability integration into Korean companies' core business and organisations is going on. Since Korean multinational enterprises (MNEs) are operating in both the Korean and the international markets, compliance with local and international environmental regulations and social standards and guidelines are very important to their strategies. As a result,

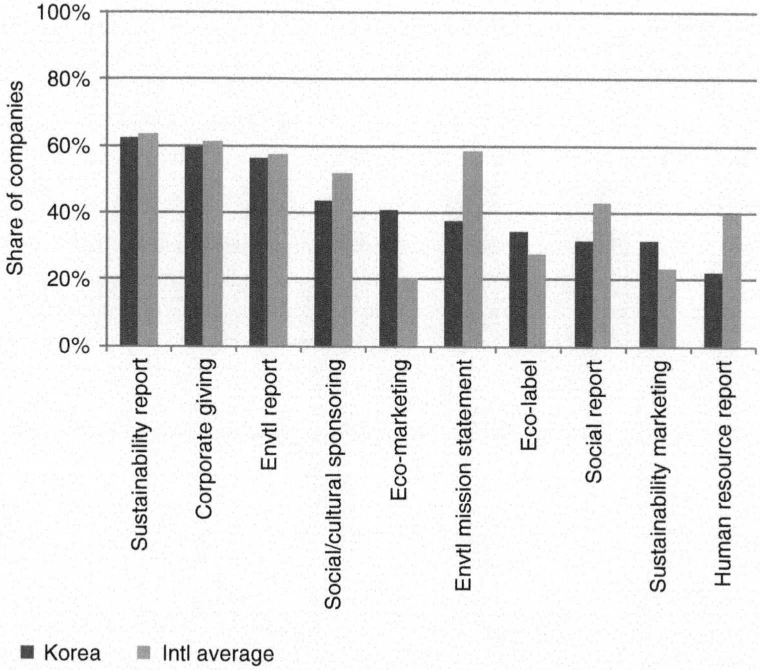

Fig. 10.22 Application of communication and marketing tools ($n_{KOR} = 32$, n_{INT} ranging from 467 to 468)

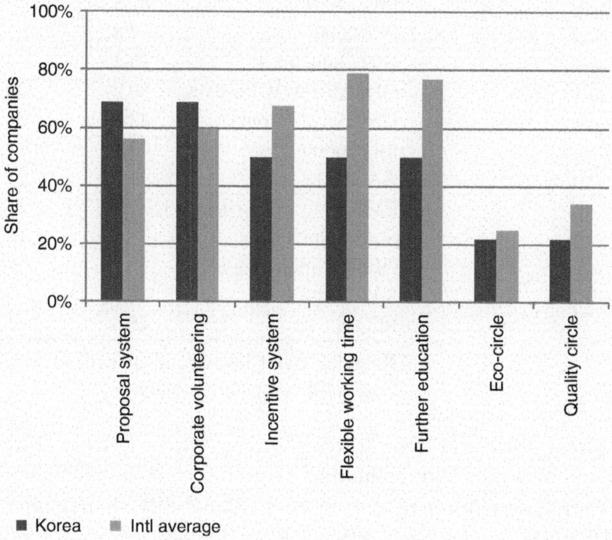

Fig. 10.23 Application of employee motivation and involvement tools ($n_{KOR} = 32$, $n_{INT} = 467$)

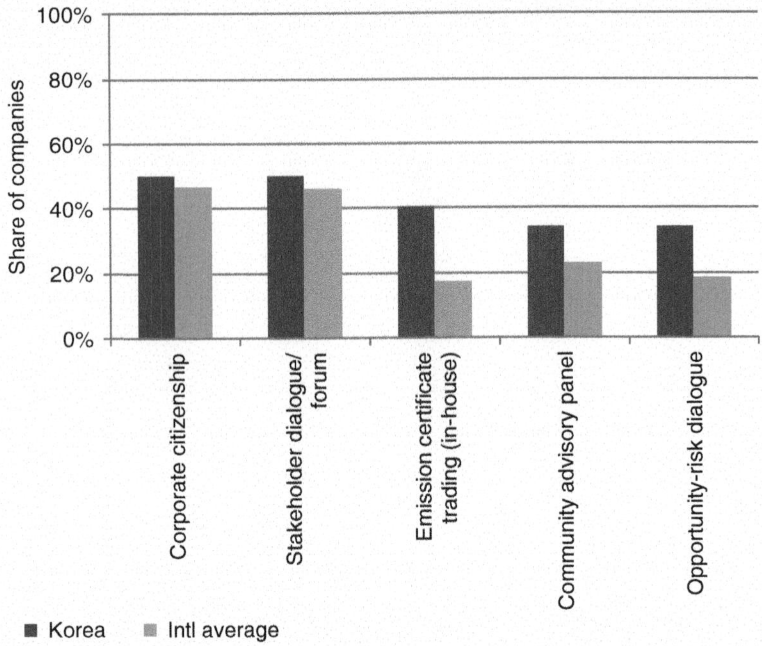

Fig. 10.24 Application of stakeholder-oriented and other tools ($n_{KOR} = 32$, n_{INT} ranging from 467 to 468)

Table 10.1 Broadly applied standards/guidelines in corporate sustainability management

Standards/guidelines	Korea (%)	International (%)
ISO 14001	81.3	73.1
ISO 9000	68.8	66.7
GRI Guidelines	53.1	52.8
OHSAS 18001/BS 8800	50.0	34.6
UN Global Compact	43.8	34.4
ISO 26000	43.8	15.4
AA 1000	37.5	12.2
EMAS	21.9	17.1
Sigma Guidelines	21.9	16.0
OECD Guidelines	18.8	18.4
SA 8000	18.8	9.8
EFQM	15.6	11.1

Note: Per cent ('known and applied' in frequency analysis) is used for comparative analysis

the involvement of a strategic planning department is a necessary step in the adoption and implementation of corporate sustainability management in Korean companies. This unique approach reflects Korea's *authoritarian chaebol business culture* involving a strategic planning department which plays an important role in adopting and implementing sustainability management. Simply put, chaebol refers to a Korean form of conglomerate which is typically controlled by a large Korean family. Once the top managers decide that corporate sustainability management is a strategically important issue, a strategic planning department or team usually develops, formulates and delivers core messages for the overall organisation to enable 'one voice and action' for an efficient and effective implementation under top management leadership.

For the implementation of corporate sustainability management, different tools and approaches are applied in Korea. Korean companies are very actively incorporating environmental management systems, quality management systems, proposal systems and corporate volunteering into their businesses. The survey of Korean companies reveals some important characteristics relating to corporate sustainability management implementation. In the recent period of rapid economic development, economic and environmental aspects of sustainability management are highly focused and developed. Many aspects of economic and environmental sustainability management are institutionally driven by the government, policy makers and standard setters. Also top managers and the strategic planning department pay a great deal of attention to the implementation of both economic and environmental sustainability management. Obviously, key stakeholders and institutional factors are two main areas for the adoption and implementation of corporate sustainability management.

In summary, there are general similarities in the adoption and implementation of corporate sustainability management in the Korean and the international samples. Although there are some strategic and operational differences between the Korean and the international sample companies, it can be concluded that companies in Korea and in an international context are aware of the importance of sustainability issues, and that they adopt available sustainability management tools and approaches to implement corporate sustainability management.

Acknowledgements We would like to thank the participating Korean companies for devoting their time and effort to the International Corporate Sustainability Barometer 2012 Survey. Ki-Hoon Lee would like to acknowledge financial support from Griffith University (International Travel Fellowship 2013 and NRG 2011/12).

References

Cheng B, Ioannou I, Serafeim G (2014) Corporate social responsibility and access to finance. Strateg Manag J 35(1):1–23

Committee GG (2009) Green growth national strategy. Green Growth Committee, Seoul

Eccles R, Serafeim G (2013) The performance frontier. Harv Bus Rev 91(5):50–60

FKI (The Federation of Korean Industries) (2011) Corporate community relations. White book. FKI, Seoul

Hoffman AJ, Woody JG (2008) Climate change: what's your business strategy? Harvard Business School, Boston

IPCC (Intergovernmental Panel on Climate Change) (2007) Climate change 2007: mitigation of climate change. Working group III contribution to the fourth assessment report of the intergovernmental panel on climate change. Cambridge University Press, Cambridge/New York

Lacy P, Cooper T, Hayward R, Neuberger L (2010) A new era of sustainability. CEO reflections on progress to date, challenges ahead and the impact of the journey toward a sustainable economy. http://www.accenture.com/SiteCollectionDocuments/PDF/Accenture_A_New_Era_of_Sustainability_CEO_Study.pdf. Accessed 8 May 2014

Lee K-H (2012a) Editorial for the special issue 'business case for sustainability: rhetoric or reality?'. Int J Innov Sustainable Dev 6(2):91–94

Lee K-H (2012b) Linking stakeholders and corporate reputation towards corporate sustainability. Int J Innov Sustainable Dev 6(2):219–235

Lee K-H (2013) Green supply chain management: implications for SMEs. In: Ndubisi N, Nwankwo S (eds) Enterprise development in SMEs and entrepreneurial firms. IGI Global, Hershey, pp 197–213

Lee K-H, Kim C-H (2014) Corporate social responsibility (CSR) practice and implementation within the institutional context: the case of the Republic of Korea. In: Low P, Idowu S, Liong A (eds) Corporate social responsibility in Asia: practice and experience. Springer, Amsterdam, pp 65–82

McKinsey (2008) How companies think about climate change: a McKinsey global survey. McKinsey Quarterly, February 2008

OECD (Organisation for Economic Co-operation and Development) (2012) OECD Korea survey. OECD, Paris

Porter M, Kramer MR (2006) Strategy & Society: the link between competitive advantage and corporate social responsibility. Harv Bus Rev 84(12):78–92

Porter M, Kramer MR (2011) Creating shared value. Harv Bus Rev 89(1/2):62–77

Porter M, van der Linde C (1995) Toward a new conception of the environment-competitiveness relationship. J Econ Perspect 9(4):97–118

Schaltegger S, Lüdeke-Freund F, Hansen E (2012) Business cases for sustainability. The role of business model innovation for corporate sustainability. Int J Innov Sustainable Dev 6(2):95–119

Schaltegger S, Harms D, Hörisch J, Windolph SE, Burritt R, Carter A, Truran S, Crutzen N, Ben Rhouma A, Csutora M, Tabi A, Kokubu K, Kitada H, Haider MB, Kim J-D, Lee K-H, Moneva JM, Ortas E, Álvarez-Etxeberria I, Daub C-H, Schmidt J, Herzig C, Morelli J (2013) International corporate sustainability barometer: a comparative study of 11 countries. Centre for Sustainability Management, Lüneburg

SGERC (Samsung Global Environment Research Center) (2012) Biodiversity conservation and role of corporations. SGERC, Seoul

Chapter 11
Exploring Sustainability in Spanish Companies

José M. Moneva, Eduardo Ortas, and Igor Álvarez

Abstract This chapter focuses on the analysis of the International Corporate Sustainability Barometer (ICSB) results in Spain, which is the fourteenth largest economy in the world by nominal GDP. A total of 23 Spanish listed companies have participated in the survey, representing about 4.9 % of the total number of companies involved in the ICSB project. Although the results show that Spanish companies carry out similar sustainability practices to those observed on international average, there are three main differences that should be highlighted. Firstly, environmental aspects are the main sustainability issues for Spanish companies. Secondly, a strong link is observed between sustainability issues and the core business of the Spanish companies. Thirdly, the main driver for the corporate social responsibility and sustainability performance of Spanish companies seems to be the legitimation of their activities. Evidence for this can be found with regard to corporate motivation (e.g. the strong influence of society-oriented stakeholders), implementation of sustainability management issues (which is strongly related to stakeholder demands) and use of sustainability reporting tools which are widely implemented by Spanish companies.

J.M. Moneva (✉) • E. Ortas
University of Zaragoza, Zaragoza, Spain
e-mail: jmmoneva@unizar.es; edortas@unizar.es

I. Álvarez
Basque Country University, San Sebastián, Spain
e-mail: igor.alvarez@ehu.es

S. Schaltegger et al. (eds.), *Corporate Sustainability in International Comparison*, Eco-Efficiency in Industry and Science 31, DOI 10.1007/978-3-319-06227-3_11,
© Springer International Publishing Switzerland 2014

11.1 Introduction

11.1.1 The Spanish Context

Spain is a member of the European Union (EU), the United Nations (UN), North Atlantic Treaty Organization (NATO), Organisation for Economic Co-operation and Development (OECD) and World Trade Organization (WTO). Spain is located in South-Western Europe on the Iberian Peninsula, and has a population of 46,818,200 people. It is a democracy organised in the form of a parliamentary government under a constitutional monarchy. The Spanish state consists of 17 autonomous regions, with 50 provinces, and 2 autonomous cities.

Spain is a mixed capitalist economy and the fourteenth-largest economy in the world by nominal GDP (255,332 million euros; IMF 2013), the fifth-largest in the EU and the Eurozone's fourth-largest. The financial crisis has had important impacts on the economy (in 2012 the GDP decreased 1.4 % and the domestic demand decreased 3.4 %), but the worst effect has been the increased unemployment rate, which has risen to more than 27 % (Banco de España 2013). Spanish industry is mainly characterised by two factors (Pérez García 2011). First, the proportion of micro-enterprises is very high (see Fig. 11.1); for some authors this situation is considered to be a main contributor to the weakness of the Spanish economy

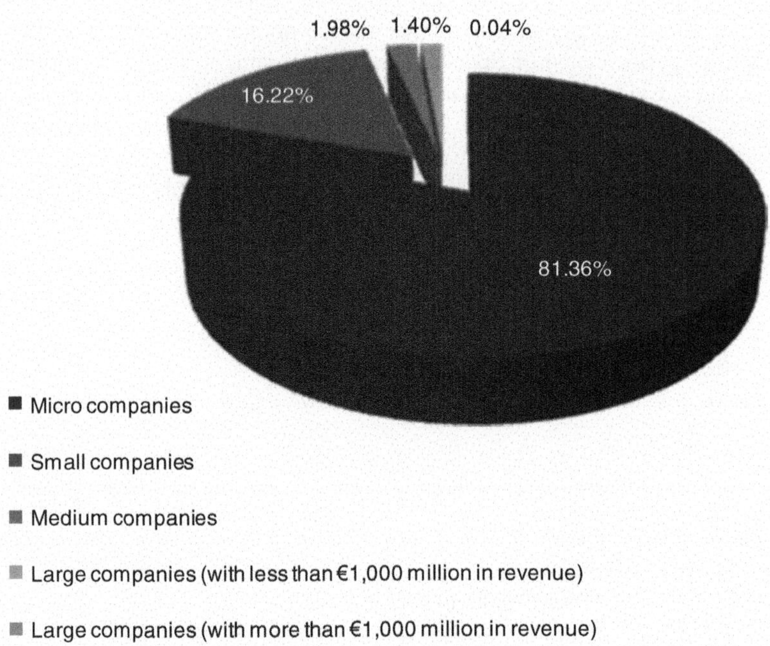

Fig. 11.1 Spanish firms by size 2010 (Adapted from Maroto 2011)

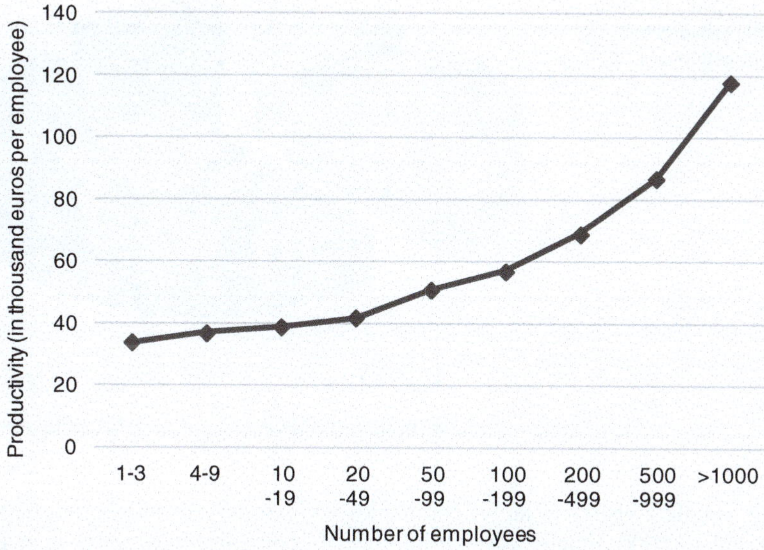

Fig. 11.2 Spanish productivity (thousand euros per employee) by company size (Adapted from PwC 2012)

(see Fig. 11.2), resulting from a high degree of competition (PwC 2012). Second, a number of Spanish holding companies are very competitive in their economic sectors worldwide. Their performance depends more on the international market than on the national business scheme. Finally, it is worth mentioning that there is a correlation between these globalised companies and success in terms of size and productivity in the Spanish context (Pérez García 2011; PwC 2012).

11.1.2 The Spanish Sample

This chapter focuses on the analysis of the International Corporate Sustainability Barometer (ICSB) results in Spain coordinated by the Centre for Sustainability Management (CSM) in Lüneburg, Germany. An overview of the results can also be found in the International Corporate Sustainability Barometer report (Schaltegger et al. 2013). Key factors regarding sustainability in Spanish companies and in-depth comparisons with the international sample of all participant countries are discussed. A total of 23 companies participated in the Spanish survey, representing approximately 4.9 % of the companies involved in the ICSB project. While this international survey focuses on large companies, describing the financial characteristics of the Spanish companies allows a better understanding of the results in the following sections. About half of the Spanish sample (47.8 %) consists of companies with employees numbering between 1,001 and 10,000, while about 26 % of the

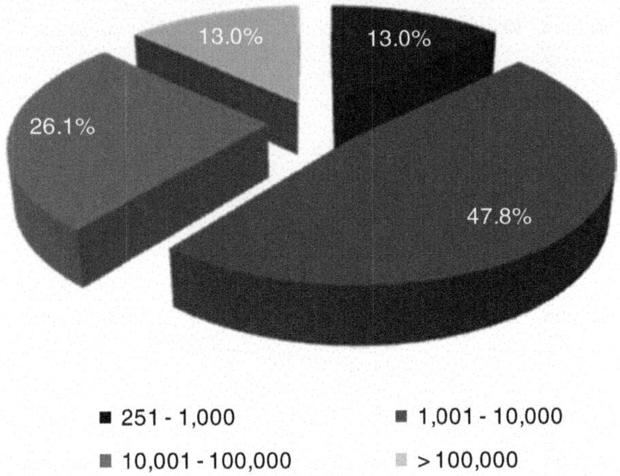

Fig. 11.3 Spanish firms in the sample grouped by number of employees ($n_{ESP} = 23$)

Spanish sample consists of larger companies with employees numbering between 10,001 and 100,000. Finally, a small number of companies have fewer than 1,001 employees or more than 100,000 (13 % in both cases; see Fig. 11.3). In spite of the differences described, all of the companies surveyed can be classified as large companies from a global perspective. The profile of the Spanish sample is very similar compared with the international sample, given that most companies in the international sample employ 1,001–100,000 employees, representing approximately 79.6 % of the international sample. However, the Spanish survey differs from the international survey in that it does not cover any companies with fewer than 251 employees. Yet, the percentage of such companies in the international sample is quite low (about 3.2 %).

Although there are some similarities between the number of employees in the Spanish and the international samples, company revenues are an issue as the Spanish and international samples differ slightly. Specifically, the Spanish companies present a high degree of asymmetry when organised according to their revenues (see Fig. 11.4). This could be because about 65 % of the Spanish companies earned up to 1,500 million euros. Furthermore, there are a limited number of companies (about 9 %) with revenues varying from more than 1,500–5,000 million euros. This situation differs from the international sample, which does not show such a concentration of firms, thus being more heterogeneous.

Moreover, there are important differences in the share of non-domestic sales (in terms of total revenue) in the Spanish and the international samples. The greatest differences are found in the first category ('more than 80 %'), where the percentage of Spanish companies is nearly half of the average sample. Only in the second category ('more than 60 % and up to 80 %') is the percentage of Spanish companies significantly higher than in the international sample. In the other categories the percentages are similar. There are no companies in the Spanish sample in the third category ('more than 20 % up to 40 %'). For this reason, Fig. 11.5 aggregates two categories in order to be comparative (i.e. 'more than 20 % and up to 60 %').

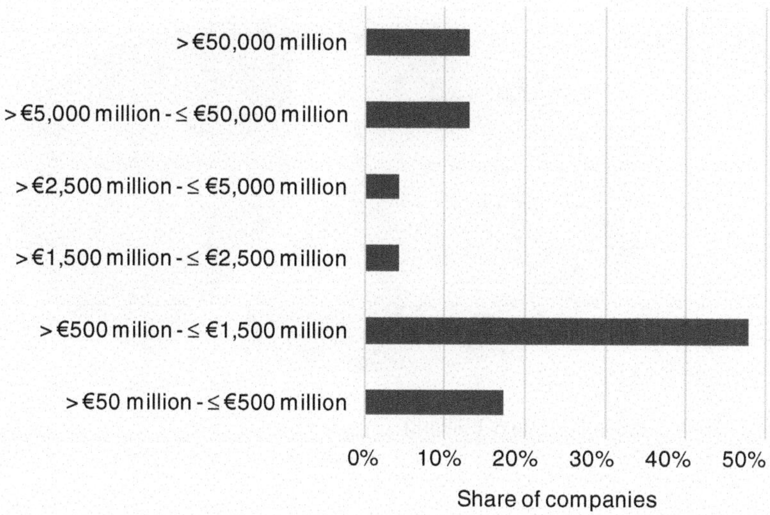

Fig. 11.4 Distribution of Spanish firms in the sample by revenue (million euros) (per cent; $n_{ESP} = 23$)

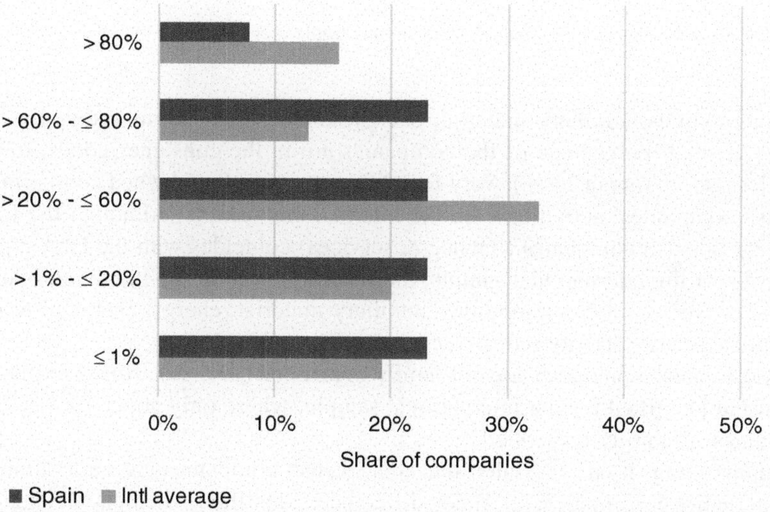

Fig. 11.5 Share of non-domestic sales in the Spanish and international samples (per cent; $n_{ESP} = 13$, $n_{INT} = 344$)

Furthermore, there is an important difference in the percentage for the 'do not know' category, since 34.8 % of the Spanish companies ticked this box (compared to 13.2 % in the international sample).

In terms of core business, most of the Spanish companies (39.1 %) are in the finance & services sector (see Fig. 11.6). Over 30 % of the companies are involved in the industry, capital goods & building sectors. Approximately 17 % are in

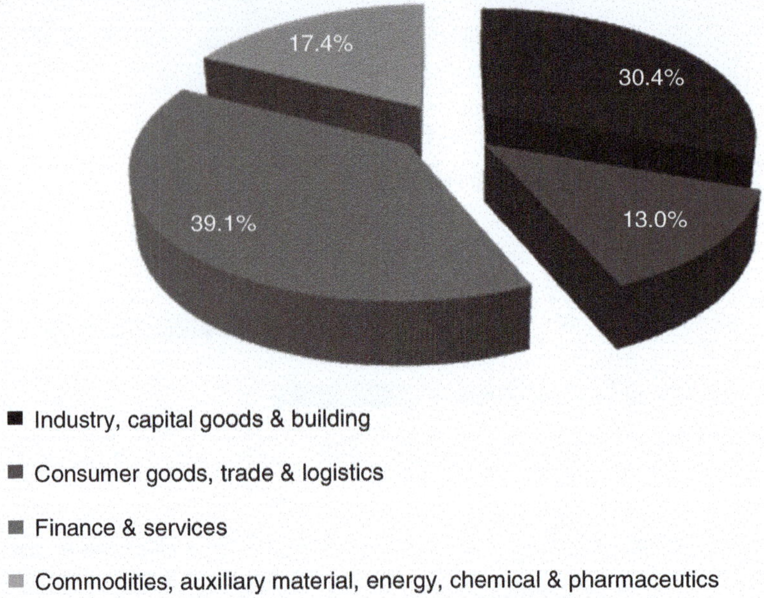

Fig. 11.6 Core business of the Spanish firms in the sample (per cent; $n_{ESP} = 23$)

the commodities, auxiliary material, energy, chemical & pharmaceutical sectors. Finally, a small percentage of the companies are in the consumer goods, trade & logistics sectors (about 13 %). Very few differences are found when comparing the Spanish companies' activities with the international sample. Although the largest group of Spanish companies (finance & services) coincides with the largest group observed for the international sample, the second largest group in the international sample consists of the commodities, auxiliary material, energy, chemical & pharmaceutical sectors and not industry, capital goods & building.

Most Spanish companies are not family-controlled (73.9 %). This is in line with the profile identified for the international sample, where only about 19.3 % of the companies are family-controlled.

Summarizing, it can be stated that the Spanish companies are quite similar to the international sample, with minor differences identified when classifying them by economic sectors and revenues.

11.2 Analysis: Sustainability Management in Spain

Sustainability has become a core strategic concern of global companies over the past two decades (Porter and van der Linde 1995; Schaltegger and Synnestvedt 2002; Rao and Holt 2005; Moneva and Ortas 2008, 2010). However, there are a number of institutional, socio-cultural and other factors that influence the sustainability practices of these companies (Ortas et al. 2012). Most of them can be classified according to country-specific factors (e.g., a country's degree of development, market regulation policies, degree of social inequalities and degree of government intervention, among others). Accordingly, this section introduces the main sustainability management practices implemented by Spanish companies. More specifically, the following topics are described: (a) intentions and motivations to implement sustainability practices; (b) sustainability integration processes; and, finally, (c) the most important tools applied to implement sustainability issues.

11.2.1 Intention

11.2.1.1 Motivation

The intentions and commitment of a company to implement sustainability practices is of utmost importance to understanding how a company integrates sustainability into its organisation. This can be explained by the fact that, depending on a company's specific motivations, different sustainability issues are addressed (Freeman 1984). This implies that companies should integrate and manage the environmental, social and economic issues, essential for their stakeholders, if they are to become successful (Freeman et al. 2010). While there are different stakeholder classification systems, the most usual models differentiate between internal and external stakeholders (Verdeyen et al. 2004).

The influence of each stakeholder on a company implementing corporate sustainability policies is rated using a scale, with a rating of 1 indicating that a stakeholder inhibits the implementation of several sustainability issues and a rating of 5 that a stakeholder is considered to promote the company's commitment to sustainability practices. A very relevant finding for the Spanish sample is that all stakeholders have a mean value greater than 3, revealing that stakeholders positively influence the implementation of sustainability practices. This is in line with the results observed in the international survey, indicating a similar influence of the various stakeholders in promoting sustainability practices in the companies. Furthermore, this analysis reveals the following relevant issues:

(a) Non-governmental Organizations (NGOs), international authorities and communities are the stakeholders considered to most strongly promote the implementation of corporate sustainability in the Spanish companies.
(b) Both the Spanish and the international samples rate international authorities higher than national authorities. Although national authorities may have greater impact than international ones, this result could be explained by the companies conception of sustainability as a global issue. This result can be also related to the fact that national legislation on the various sustainability topics is soft and has little effect on the Spanish companies (Llena et al. 2007).

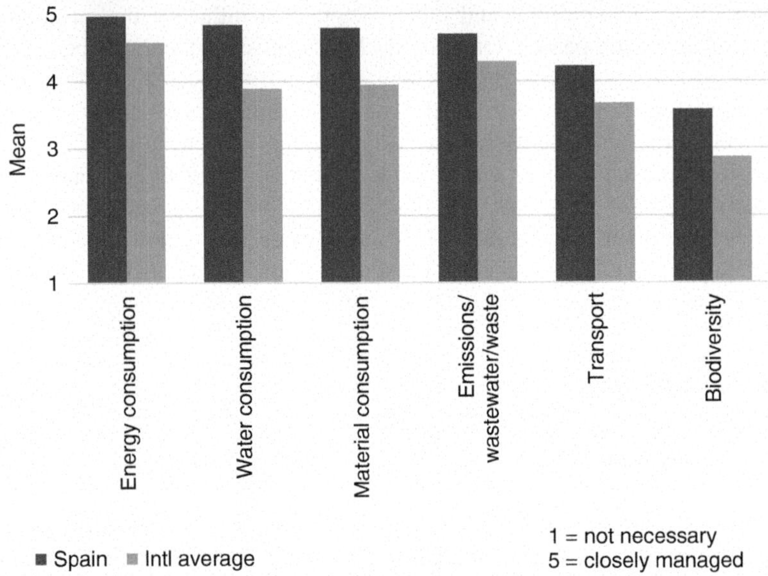

Fig. 11.7 Managed environmental issues ($n_{ESP} = 23$, n_{INT} ranging from 455 to 463)

(c) Insurance companies and trade unions are rated as the stakeholders least promoting sustainability in both the Spanish and the international samples. The literature establishes that one of the main factors in improving a company's sustainability performance is the risk related to penalties for environmental or social damage, and the stakeholders most concerned with this risk are financial institutions. It can be observed for those stakeholders in the Spanish sample, however, that this issue is less relevant than for the others, such as NGOs and the community. The most important differences between the Spanish and the international samples appear in the analysis of the stakeholders with the least degree of influence in establishing sustainability practices. In fact, rating agencies are a relevant stakeholder in the international sample (with a mean value of 3.53). However, these stakeholders reach the third lowest score in the Spanish context, similar to that obtained by banks (both 3.29). This difference is quite interesting because rating agencies strongly impact financial markets and most companies are attentive to their recommendations.

11.2.1.2 Issues

Another important aspect that enhances or inhibits the consideration of sustainability in a firm's strategy is the kind of sustainability issues a company manages. To better understand this question, social and environmental sustainability issues are analysed separately (see Figs. 11.7 and 11.8). A preliminary overview of the

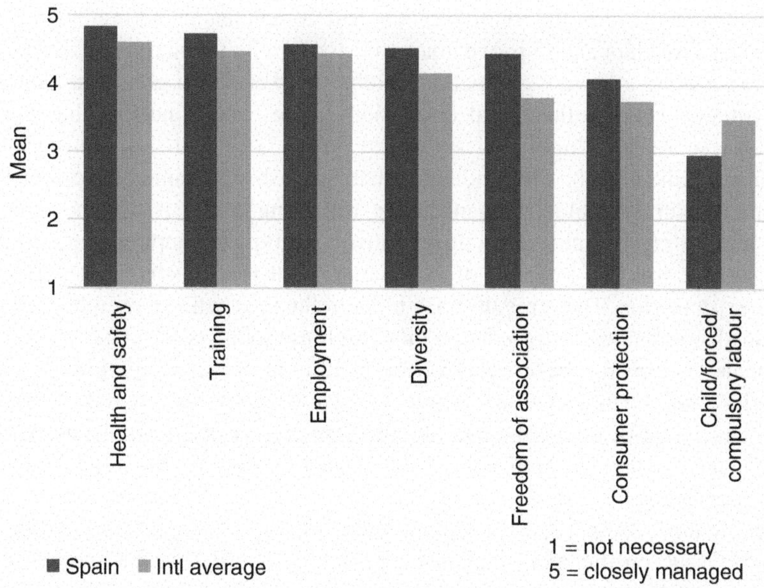

Fig. 11.8 Managed social issues (n_{ESP} ranging from 22 to 23, n_{INT} ranging from 442 to 461)

results indicates that for the Spanish companies the environmental issues are of more interest than the social ones. It is quite difficult to provide an accurate reason for this result in the Spanish context without conducting a more in-depth analysis of the social and environmental issues managed by the Spanish companies and the internal and external factors shaping their choices. However, a driver that may explain this result is the impact of environmental disasters on Spanish society over the last several decades (e.g. Aznalcollar dam failure, the Prestige oil spill, uncontrolled forest fires every summer and water scarcity in dry regions, among others).

In the international sample of companies the social issues are more closely managed than the environmental ones. A more in-depth analysis reveals an interesting finding, which is the fact that Spanish companies score a median value of 5 concerning the four environmental issues managed most closely. This clearly indicates that Spanish companies have developed measures to closely manage aspects related to energy consumption, water consumption and materials consumption, as well as emissions, wastewater and waste. On the other hand, biodiversity and transportation are the environmental issues less closely managed. However, the mean score of Spanish companies with regards to these two environmental issues is significantly higher than the international average (about 3.57 and 4.22, respectively). The differences between Spain and the international sample are then quite significant. All environmental issues are managed more closely in the Spanish sample than the international average (difference between 0.39 and 0.95). The main differences are observed with regards to water consumption. This situation could be influenced by the recurring water shortages in some Spanish regions.

These findings are in line with the view that companies operating in developed markets need to adapt and improve innovative tools to better manage environmental issues in an open and very competitive global market. What about the degree of management related to the social dimensions? The results indicate that Spanish companies score very highly on six social issues (i.e., with a median score of 5 for these dimensions). This indicates that Spanish companies have developed different measures to closely manage the following social issues: occupational health and safety, training and development, workplace/employment, diversity and equal opportunity, freedom of association/right to collective bargaining and consumer protection. These results are similar to the international sample. However, the Spanish companies perform better on all of these social issues. One exception is that Spanish companies have not developed essential processes to closely manage a high risk topic for transnational companies: child labour, forced and compulsory labour. This could be due to the fact that this issue rarely occurs on a national level or that Spanish companies make only limited use of suppliers from regions where it is most prevalent.

Thus, Spanish companies should continue working on recognising the impact of this social issue on the supply chains, especially when the production process is totally or partially outsourced to developing countries (Vachon and Klassen 2006), which in general take little account of ethical standards in business (Ortas et al. 2012).

11.2.2 Integrating Sustainability

11.2.2.1 Connection to Core Business

This section analyses whether the Spanish companies are able to integrate the environmental and social aspects of sustainability into their core business. This integration process is pivotal for companies when adopting sustainability-oriented strategies. The results of the survey show that 43.5 % of Spanish companies consistently connect sustainability to all segments of their core business (see Fig. 11.9). Furthermore, 39.1 % of the Spanish companies connect sustainability issues to most segments of their core business. It is highly relevant that only 17.4 % of the companies indicate they connect sustainability to some or only a few segments of their core business. Furthermore, when comparing these results with those obtained for the international sample, a superior performance on the part of Spanish companies with regards to this issue is revealed. These findings indicate that the integration of environmental and social policies into the core business is higher in Spanish companies than the average of the international sample.

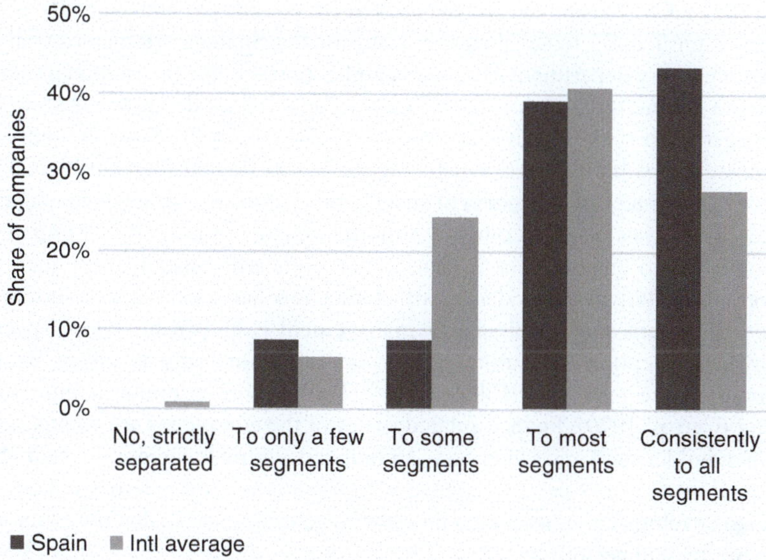

Fig. 11.9 Sustainability integration into core business (per cent; $n_{ESP} = 23$, $n_{INT} = 457$)

11.2.2.2 Involvement of Organisational Units

In addition to the integration level of sustainability processes, an analysis of the relationship of these activities with organisational units/internal stakeholders is necessary to provide a better overview of the implementation of sustainability aspects into core business processes. This is of special importance because in the Stakeholder Theory (Freeman 1984), a reporting-driven sustainability accounting development process can be initiated on the basis of a stakeholder or shareholder-orientated view or a multiple stakeholder engagement process (Burritt and Schaltegger 2010). The first relevant finding is that all of the organisational units analysed in the Spanish survey are well engaged in sustainability-related measures. This is shown by the fact that the mean value associated with each one of the internal stakeholders is equal to or greater than 3. The only exception is observed for accounting. However, this unit achieved a mean value close to 3 (2.95). Furthermore, the corporate social responsibility (CSR)/sustainability department, top management and human resources (HR) department are the most engaged with sustainability measures (i.e., their mean values are 4.96, 4.23 and 4.23, respectively).

The least engaged organisational units are accounting, finance as well as logistics and distribution. However, these organisational units are, on average, not perceived as inhibiting but are assessed as more neutral. These findings are similar to the pattern observed in the international sample, even though the international scores for the internal stakeholders ororganisational units evaluated as most strongly engaged (i.e., CSR/sustainability department, top management and HR department) are significantly lower than the score for the Spanish companies (only the CSR/sustainability department has a mean higher than 4 in the international sample). Also, in the international sample the accounting and finance departments seem to be less engaged with sustainability measures than they are in the Spanish companies. These positive results for the Spanish companies can be considered a logical consequence of the integration process of sustainability into core strategic management activities. This is because if a company aims to implement sustainability into its core business, it has to engage most organisational units with sustainability measures. However, the relatively low degree of connection between the accounting and finance department and sustainability management is surprising. This aspect fits in with the perception that, although corporate sustainability is linked with the triple bottom line approach (Elkington 1997), therefore adopting a multivariate character (Carroll 1979), finance-related issues of the companies are mostly related to the shareholders and other investors. A company aiming to integrate sustainability strategies does not only need to engage with its internal and external stakeholders but also to develop key indicators to monitor its activities and to determine its social and environmental performance.

11.2.3 Implementing Sustainability with Sustainability Management Tools

This section focuses on the implementation of sustainability into business management. This will contribute to obtaining a better picture of how sustainability management tools are applied by the Spanish companies. To this end 79 sustainability management tools were investigated to determine which of them are the most implemented by the Spanish companies. The main results indicate that corporate citizenship (78.3 %), environmental management systems (EMS; 73.9 %), sustainability reports (73.9 %), flexible working time (73.9 %), and further education (73.9 %) are the most frequently implemented sustainability methods in Spanish companies. The outstanding difference to the international average is related to the corporate citizenship tool, which is applied by 46.7 % of the companies in the international sample. This can be explained by the fact that large Spanish companies support foundations which mainly focus on giving financial support and developing social and environmental activities/initiatives.

Another important finding is that sustainability reporting seems to be of less importance for the international sample than for the Spanish one. This could be a result of the high degree of acceptance of these non-conventional reports by both Spanish companies and their stakeholders. This aspect has been commented on in recent academic literature in the field, which indicates that Spain is the country with the greatest number of companies publishing Global Reporting Initiative (GRI) reports worldwide (Álvarez 2010). Looking at other sustainability tools in Spanish companies, it can be seen that eco/social checklists (8.7 %), social/fair labelling

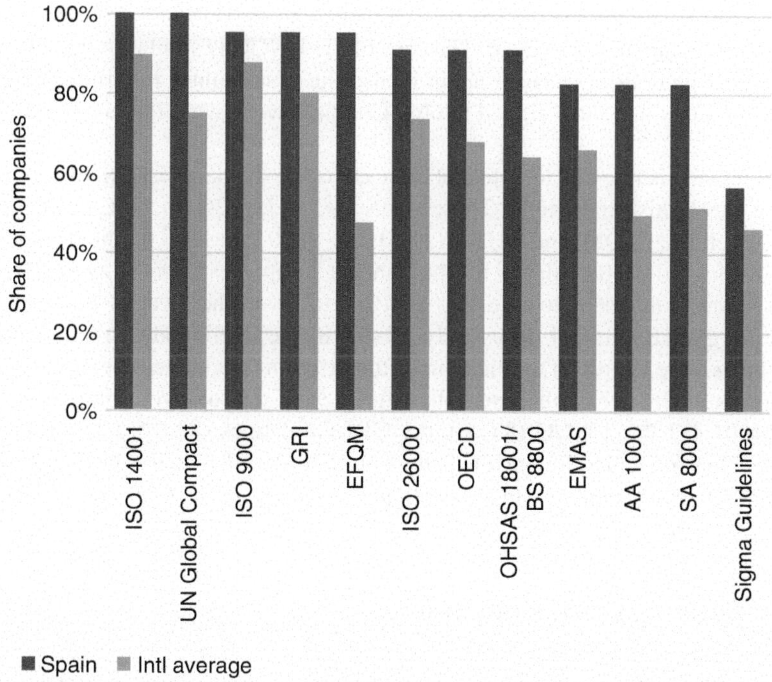

Fig. 11.10 Awareness of sustainability standards/guidelines (per cent; $n_{ESP} = 23$, $n_{INT} = 457$)

(8.7 %), sustainability labels (8.7 %) community advisory panels (13 %), eco-labels (13 %), eco-marketing (13 %), eco-budgeting (13 %), sustainability accounting (13 %), eco-sponsoring (13 %), eco/social ABC analysis (13 %) and environmental shareholder value (13 %) are the least implemented. It is interesting to note that among the least applied tools can be found those related to customers/marketing (e.g., social/fair labelling, sustainability labels, eco-labels, eco-marketing). This can be due to the fact that Spanish companies perceive that a customer perspective is not sufficiently developed in their domestic markets. This situation is not observed in the international sample, where those tools achieve higher scores.

An additional surprising or even peculiar result is the low rating for sustainability accounting, because as we observed above the application of GRI sustainable reporting by the Spanish sample is very high. This situation may be linked to the problems of compulsory environmental disclosures implementation in financial statements (Llena et al. 2007).

Another relevant issue that can explain the degree of sustainability implementation in a specific company refers to the level of knowledge and application of the different international standards or guidelines (Ortas and Moneva 2011). In order to investigate this issue, companies were surveyed about their knowledge and application of 12 international standards/guidelines. The results indicate that nearly all standards are known by all the Spanish companies (see Fig. 11.10). Specifically, the ISO 14001 and the United Nations Global Compact (UNGC) are known by all the Spanish companies surveyed. The case of the UNGC is similar to that observed for the GRI guidelines because Spain is also the first country in terms of number of organisations that have signed Global Compact worldwide (Garayar and Calvo 2012).

Moreover, it is relevant to note that there are other standards known by more than 90 % of the Spanish companies. These are ISO 9000, ISO 26000, OECD Guidelines, GRI Guidelines, EFQM, and OHSAS 18001/BS 8800. The level of knowledge of all standards investigated is higher for the Spanish companies than the average results of the international sample (see Fig. 11.10). However, the level of knowledge of a specific guideline is not necessarily related to the likelihood of its application. Nonetheless, the degree of application of the different guidelines is extremely high in Spain as well. Specifically, more than the 70 % of the surveyed companies apply ISO 14001 (87 %), UNGC (87 %), the GRI Guidelines (82.6 %) and ISO 9000 (73.9 %). In sum, and following the pattern already identified in this project, Spanish companies perform better on this issue than the average of the total sample of firms.

11.3 Summary and Conclusions

The International Corporate Sustainability Barometer analyses the state of corporate sustainability practice in eleven different countries around the world. This analysis reveals that managing sustainability is a global challenge and opportunity that large companies all over the world have in common. Spanish companies have included sustainability in their corporate agendas in ways similar to companies from other industrialised countries.

The main driver for this behaviour in Spain is the prevalence of society-oriented stakeholders, specifically NGOs and environmental/social organisations. The relevance of banks is low, but higher than in most of the other countries surveyed, indicating an above average impact of the financial sector on companies.

Management levels of aspects of sustainability in Spanish companies are high. This is obvious in the findings regarding the integration of sustainability into core business, where most of the Spanish companies report they consistently connect sustainability to all segments of the core businesses. The units most engaged with sustainability are been closely managed by Spanish companies: child labour, forced and compulsory labour, which reveals a limited analysis of sustainability in the supply chain. These findings may be connected with the low share of non-domestic sales in the Spanish context.

Moreover, the prevailing use of measurement tools based on quality (e.g. environmental and quality management systems) and reporting (e.g. sustainability reporting) may validate the legitimacy of the approach of Spanish companies with regards to sustainability issues. A reason may be that environmental issues, which usually have measures that are generally accepted, are more attractive than social issues.

Finally, the ICSB indicates that Spanish companies have similar levels of sustainability performance to other large companies around the world. They have certain strengths, such as the linking of sustainability to the core business and the use of sustainability management tools. However, they might consider increasing stakeholder engagement as a means of effectively managing sustainability issues as well as improving the analysis of the impact of social issues, especially in the supply chain.

Acknowledgements The authors are grateful for the financial help provided by the Spanish Ministry of Science and Innovation (research project ECO 2011-26171). The usual disclaimer applies.

References

Álvarez I (2010) Sustainability reports as a tool to manage sustainability. Span J Finance Account 38(144):677–697

Banco de España (2013) Boletín económico, marzo 2013: Evolución reciente de la economía española. http://www.bde.es/f/webbde/SES/Secciones/Publicaciones/InformesBoletinesRevistas/BoletinEconomico/13/Mar/Fich/be1303-evo.pdf. Accessed 20 June 2013

Burritt RL, Schaltegger S (2010) Sustainability accounting and reporting. Fad or trend? Account Audit Account J 23(7):829–846

Carroll AB (1979) A three-dimensional conceptual model of corporate performance. Acad Manage Rev 4(4):497–505

Elkington J (1997) Cannibals with forks: the triple bottom line of 21st century business. Capstone Publishing, Oxford

Freeman RE (1984) Strategic management: a stakeholder approach. Prentice Hall, Englewood Cliffs

Freeman RE, Harrison JS, Wicks AC, Parmar BL, De Colle S (2010) Stakeholder theory: the state of the art. Cambridge University Press, Cambridge

Garayar A, Calvo JA (2012) Joining the UN global compact in Spain: an institutional approach. Span Account Rev 15(2):311–355

IMF (International Monetary Fund) (2013) Article IV Consultation with Spain—Concluding Statement of the Mission. http://www.imf.org/external/np/ms/2013/061813.htm. Accessed 20 June 2013

Llena F, Moneva JM, Hernandez B (2007) Environmental disclosure and compulsory accounting standards: the case of Spanish annual reports. Bus Strateg Environ 16(1):50–63

Maroto JA (2011) Las PYMES españolas en forma societaria. Colegio de Registradores de la propiedad, bienes muebles, y mercantiles de España, Madrid

Moneva JM, Ortas E (2008) Are stock markets influenced by sustainability matter? Evidence from European companies. Int J Sustain Econ 1(1):1–16

Moneva JM, Ortas E (2010) Corporate environmental and financial performance: a multivariate approach. Ind Manage Data Syst 110(2):193–210

Ortas E, Moneva JM (2011) Sustainability stock exchange indexes and investor expectations: multivariate evidence from DJSI-Stoxx. Span J Finance Account 11(151):395–416

Ortas E, Moneva JM, Salvador M (2012) Does socially responsible investment equity indexes in emerging markets pay off? Evidence from Brazil. Emerg Markets Rev 13(4):581–897

Pérez García F (2011) Informe Fundación BBVA-Ivie sobre crecimiento y competitividad, trayectorias y perspectivas de la economía española 1995–2010. Fundación BBVA-Lvie. http://www.fbbva.es/TLFU/dat/Informe%202012_FBBVA_Ivie.pdf. Accessed 19 June 2013

Porter ME, van der Linde C (1995) Toward a new conception of the environment-competitiveness relationship. J Econ Perspect 9(4):97–118

PwC (2012) Temas candentes de la industria española para 2012: Necesitamos un modelo. PwC, Madrid

Rao P, Holt D (2005) Do green supply chains lead to competitiveness and economic performance? Int J Oper Prod Manage 25(9):898–916

Schaltegger S, Synnestvedt T (2002) The link between 'green' and economic success. Environmental management as the crucial trigger between environmental and economic performance. J Environ Manage 65(4):339–346

Schaltegger S, Harms D, Hörisch J, Windolph SE, Burritt R, Carter A, Truran S, Crutzen N, Ben Rhouma A, Csutora M, Tabi A, Kokubu K, Kitada H, Haider MB, Kim J-D, Lee K-H, Moneva JM, Ortas E, Álvarez-Etxeberria I, Daub C-H, Schmidt J, Herzig C, Morelli J (2013) International Corporate Sustainability Barometer: A Comparative Study of 11 Countries. Centre for Sustainability Management, Lüneburg

Vachon S, Klassen RD (2006) Extending green practices across the supply chain. Int J Oper Prod Manage 26(7):795–821

Verdeyen V, Put J, Buggenhout V (2004) A social stakeholder model. Int J Soc Welf 13(4):325–331

Chapter 12
State of the Art and Progress of Corporate Sustainability in Switzerland

Jörg E.U. Schmidt and Claus-Heinrich Daub

Abstract The state of the art and the progress of sustainability management were surveyed among 205 companies with headquarters in the German-speaking part of Switzerland; 25 of those companies responded to the ICSB questionnaire. Compared to the international average, the sample included a larger number of small companies (between 250 and 10,000 employees) and from the finance & services sector, but fewer companies from the industry sector. For many questions, the results indicate similar trends compared to the international average, but values from Swiss companies are often lower. Some differences were noticeable. Although social and environmental issues were considered less important for the management than on international average, a substantial knowledge and relatively high level of application of management tools is striking. Also the influence of many external stakeholders is considered less promoting for the implementation of corporate sustainability. The findings are discussed in relation to the particularities of the Swiss sample.

12.1 Introduction

12.1.1 The Swiss Context

During the past years, sustainability, which in this study is defined as the integration and management of economic, environmental and social issues in corporate activities, performance and communication, has also become a key issue for many Swiss companies (Berger et al. 2012; ISDC 2012). In Switzerland, critical and

J.E.U. Schmidt (✉) • C.-H. Daub
School of Business, Institute of Management, University of Applied Sciences and Arts Northwestern Switzerland, Windisch, Switzerland
e-mail: joerg.schmidt@fhnw.ch; clausheinrich.daub@fhnw.ch

S. Schaltegger et al. (eds.), *Corporate Sustainability in International Comparison*, Eco-Efficiency in Industry and Science 31, DOI 10.1007/978-3-319-06227-3_12,
© Springer International Publishing Switzerland 2014

politically active citizens have a long tradition of asserting their needs and demands to companies and the government. Stakeholders, including the society at large, have used established instruments of direct political participation to exercise influence on economic and business issues, a recent example being restrictions on executive compensation in companies listed on the Swiss stock market and the strengthening of shareholders in corporate governance (Wagner and Wenk 2013).

In spite of its small size and population of more than eight million people, Switzerland has one of the highest per capita income rates worldwide (World Bank 2006) and is home to many multinational companies, especially in the finance (UBS, Credit Suisse), chemical (Novartis, Roche, Syngenta), food (Nestlé),mining (Holcim) and commodity trading sectors (GlencoreXstrata, Mercuria Energy Group). Other well-known areas which contribute to the global recognition and image of Switzerland are luxury goods and watches (Rolex, Swatch-Group), machinery (ABB, Sulzer) and special food products like chocolate (Barry Callebaut, Lindt & Sprüngli) or cheese. Some of the largest Swiss companies, however, are in the retail (COOP, Migros), logistics (Schweizerische Post), public transport (SBB) and communications sectors (Swisscom), and enjoy a high level of public recognition and popularity. "Swissness", the combination of well-recognised characteristics of Swiss companies, is a powerful nation brand.

Switzerland has 26 administrative entities (cantons) and is divided into four language regions (German, French, Italian and Romansh). The economic centre is in the German-speaking region, with the large cities of Zurich, Basel and Bern, and to some extent in the French-speaking region (Geneva, Lausanne). Cultural, social and political differences exist between the language regions, which are often manifested in the results of national referendums. Switzerland is a landlocked country in central Europe, bordering five countries (Germany, Liechtenstein, Austria, Italy and France). Although not itself a member of the EU, its contacts with its EU neighbours and the EU itself are intensive on a political and economic level, with EU member states as its main business partners.

Health and sustainability aspects are important for a large share of Swiss consumers, which can be seen in the increasing sales of products with labels for organic food (e.g. Bio Suisse) or social aspects (e.g. Max Havelaar). As a result Swiss companies in different sectors are actively reconsidering the role of sustainability in their management and, increasingly, in their corporate strategy.

12.1.2 The Swiss Sample

The total set of Swiss companies which were contacted and invited to participate in the ICSB study, coordinated by the Centre for Sustainability Management (CSM) in Lüneburg (see Schaltegger et al. 2013), was gathered from the list of the largest companies in Switzerland issued annually by the Swiss business newspaper 'Handelszeitung'. This list categorises companies as "industrial, commercial and service enterprises", "banks/financial institutions", and "insurance companies", and

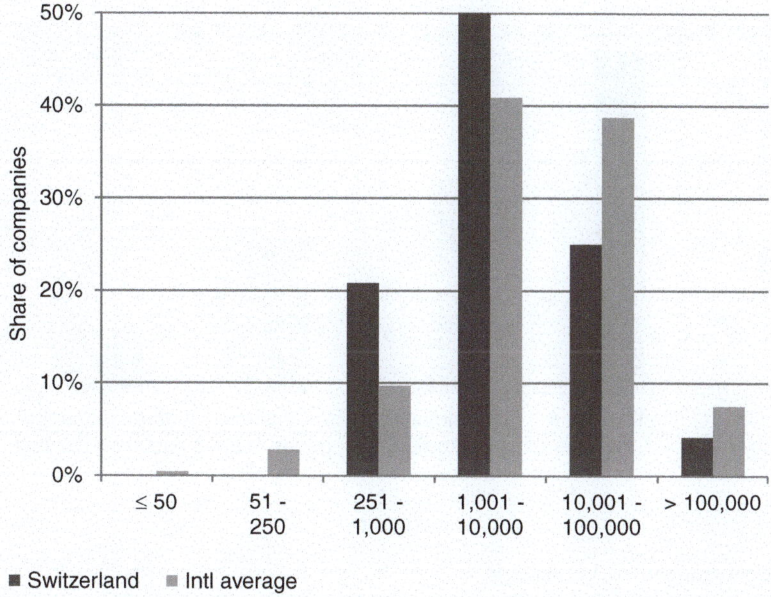

Fig. 12.1 Classification of the responding companies according to their numbers of employees (per cent; $n_{SUI} = 24$, $n_{INT} = 465$)

their size is determined and ranked on the basis of their annual revenues, including annual sales, total assets, or gross premiums, respectively. The most recent list for 2011 was used for the study (Dun & Bradstreet (Schweiz) AG 2011), based on financial results from 2010.

Companies in the list were only contacted if their financial results were over 50 million Swiss francs in 2010, if their headquarters were in the German-speaking part of Switzerland and if they were not subsidiaries of other companies on the list. A total of 205 companies were contacted and invited to participate in the study by email in June 2012. Emails were sent in German or English. Companies which had not responded in the following weeks were invited again by phone in July and August 2012.

Of the contacted Swiss companies 25 participated, yielding a response rate of 12.2 % and a share of 5.3 % of all companies participating in the ICSB study. Not all participating Swiss companies gave valid responses to all questions in the questionnaire. Per cent values given in this chapter refer to the valid responses.

According to their own classification, most participating Swiss companies have between 1,001 and 10,000 employees (50.0 % per cent of companies with valid responses) or between 10,001 and 100,000 employees (25.0 %). No small and medium-sized Swiss companies (with less than 250 employees) participated in the study. On international average, larger companies with 10,001–100,000 employees make up a considerably higher proportion (38.7 %) than in the Swiss data set (see Fig. 12.1).

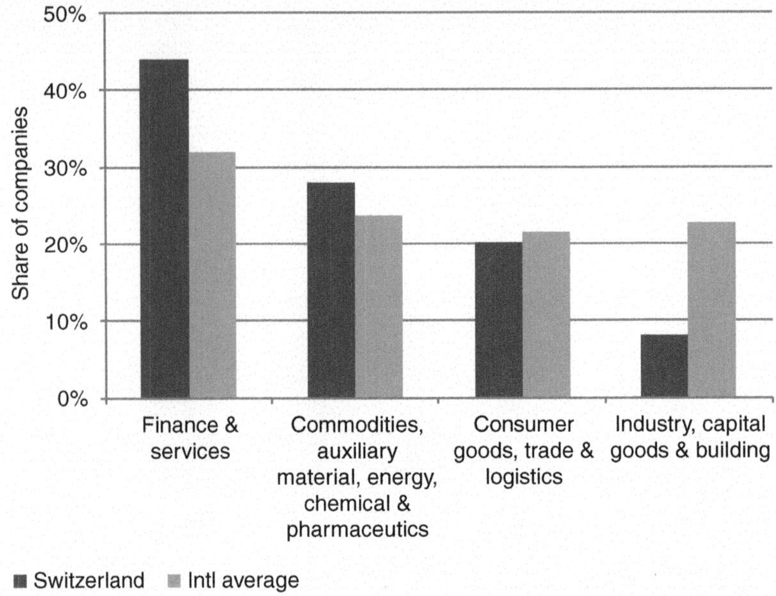

Fig. 12.2 Classification of the responding companies according to their core business (per cent; $n_{SUI} = 25$, $n_{INT} = 468$)

Furthermore, most (44.0 %) of the participating Swiss companies stated that they had earned between €5,000 and €50,000 million in revenue in the previous financial year, 28.0 % between €500 and €1,500 million. Similar proportions are represented in the international sample, but no Swiss companies with a revenue of more than €50 billion participated in the study.

Of the participating Swiss companies 44.0 % specified their core business as "finance & services", 28.0 % as "commodities, auxiliary material, energy, chemical & pharmaceutical industry", 20.0 % as "consumer goods, trade & logistics", and 8.0 % as "industry, capital goods & building". The Swiss data set therefore includes a much higher proportion of finance and service companies than the international average, which has only 32.1 % of the responding companies in this sector. Companies from the industry, capital goods & building sector make up a much higher portion (22.6 %) than in the Swiss data set (see Fig. 12.2).

A small minority of only 8.3 % of the participating companies are family-controlled, compared to the international average, where family-controlled companies make up a slightly larger proportion (19.3 %).

The Swiss data set includes a large share of companies (42.1 %) with more than 80 % of non-domestic sales in total revenue in the previous financial year and another large share (31.6 %) with only up to 1 % of non-domestic sales. This is in contrast to the international average, where companies are relatively evenly represented with regard to this characteristic.

12.2 Analysis: Switzerland in International Comparison

12.2.1 Intention

12.2.1.1 Motivation

Corporate sustainability is the result of influence by a range of stakeholders, various promoting and inhibiting factors and the potential effects which companies themselves expect from engaging in ecological and social topics.

Considering the impact of different external stakeholders on the implementation of sustainability, international authorities (mean average: 3.71) and NGOs (3.64) have the highest promoting effect for Swiss companies. At the other end of the spectrum, suppliers (3.00), banks (3.00), trade associations (2.96) and trade unions (2.86) appear to have a relatively indifferent or even inhibiting impact. Overall, the influence of almost all stakeholders (except for international authorities) is rated lower (i.e. considered neutral or even slightly inhibiting) by Swiss companies than on international average; this is particularly noticeable for NGOs, trade associations, media, the community, consumer organisations and national authorities (see Fig. 12.3).

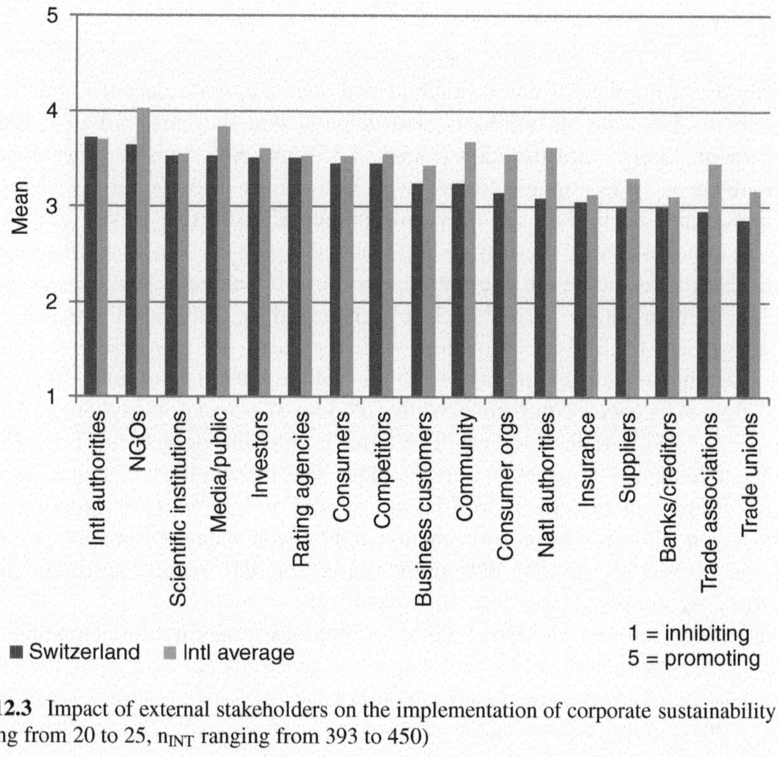

Fig. 12.3 Impact of external stakeholders on the implementation of corporate sustainability (n_{SUI} ranging from 20 to 25, n_{INT} ranging from 393 to 450)

With regard to the impact of different promoting factors on the implementation of sustainability, Swiss companies assessed know-how in the company (mean: 4.00) and commitment by top management (3.96) to have the highest, government incentives the lowest promoting impact (3.04). Almost all factors specified in the study (except for know-how in the company) have a lower promoting impact for Swiss companies than on international average. From a choice of inhibiting factors, Swiss companies assessed lack of personnel capacities (3.32) and lack of government incentives (3.24) as having the highest impact on the implementation of sustainability in their company, lack of self-commitment by business the lowest (2.39). Compared to the international average, lack of financial capacities, legislation and lack of self-commitment by business appear to have a less inhibiting impact. A comparison of promoting and inhibiting factors shows that the impact of promoting factors was considered higher than that of inhibiting factors both in the Swiss and the international samples.

Swiss companies consider the enhancement and safeguarding of corporate reputation (mean: 4.38) and employee motivation (3.92) as the most important positive impacts of corporate sustainability, business model innovation (3.08) and revenue increase (2.96) the least important. The ranking of impact factors is similar to the international average, while companies belonging to the international sample consider most of the specified impacts to be more important.

12.2.1.2 Issues

Not only are a number of environmental and social issues relevant for the companies themselves, but stakeholders also demand that they are part of corporate management. Energy consumption (mean: 4.32) and emissions/wastewater/waste (3.92) are most closely managed by Swiss companies. Biodiversity (2.56) is the least important and is thus the only environmental issue that is still considered relatively unnecessary by the responding Swiss companies. The same trend can be perceived in the results of the international average, but all issues have lower means in the Swiss data set, i.e. are less closely managed or considered less necessary for management by Swiss companies. With regard to the management of social issues, training and development (4.32) as well as health and safety (4.08) are ranked highest by Swiss companies, while freedom of association/right to collective bargaining (2.88) is ranked lowest. The latter is thus the only social issue that is considered relatively unnecessary to manage. The international average shows a slightly different picture with child labour, forced and compulsory labour ranked lowest. Again, all issues have lower means in the Swiss data set. On average, social issues are ranked higher than environmental issues with respect to the extent to which they are managed (see Figs. 12.4 and 12.5).

Stakeholder demands as perceived by companies reflect a similar picture with regard to environmental issues, but the mean averages for *managed* issues are higher than the respective means for *demanded* issues. Again, energy consumption (mean: 3.08) and emissions/wastewater/waste (2.76) are ranked highest by Swiss

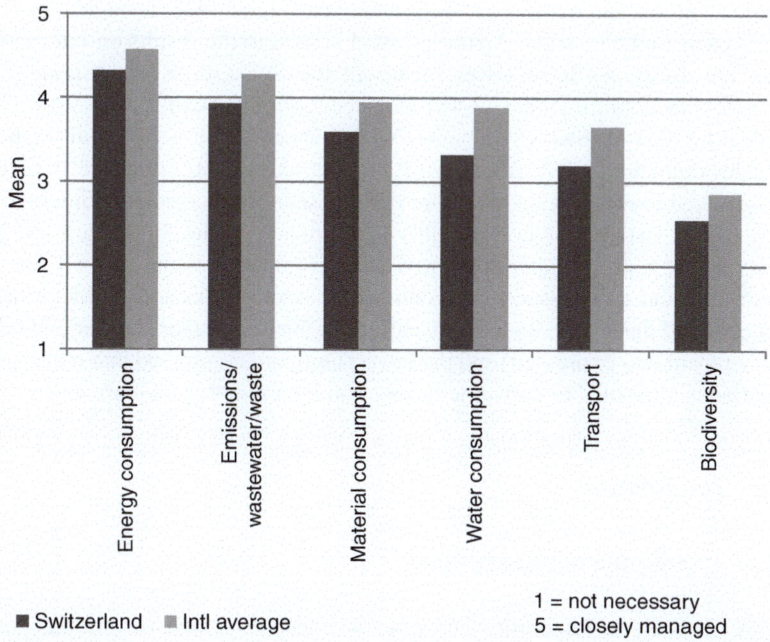

Fig. 12.4 Management of environmental issues (n_{SUI} ranging from 24 to 25, n_{INT} ranging from 455 to 463)

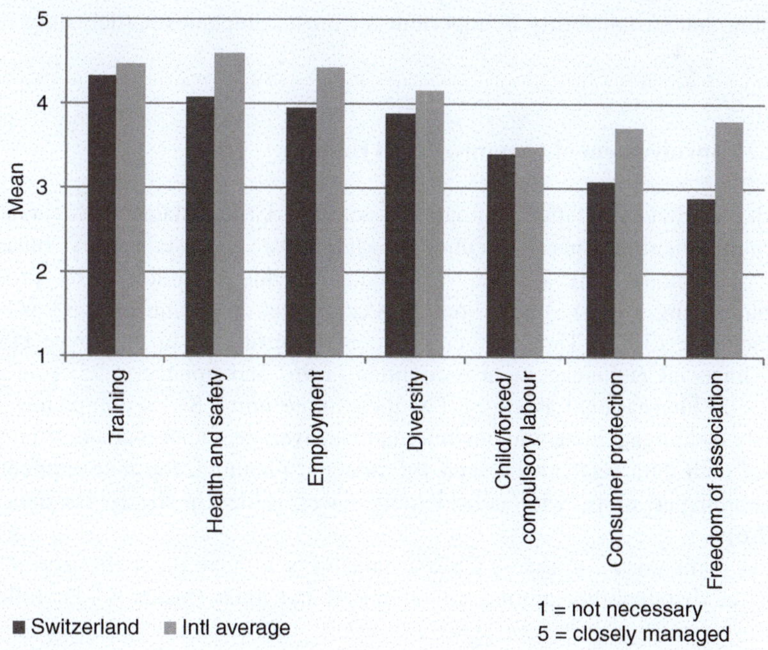

Fig. 12.5 Management of social issues (n_{SUI} ranging from 24 to 25, n_{INT} ranging from 442 to 461)

companies with respect to the issue's importance in stakeholder demands; biodiversity (2.28) was ranked lowest. A similar trend is seen in the results on international average, but all issues have lower means in the Swiss data set. Among social issues, workplace/employment (3.48) and occupational health and safety (3.36) are ranked highest by Swiss companies with respect to the issue's importance in stakeholder demands, while freedom of association/right to collective bargaining (2.16) is ranked lowest. Again, a similar trend is seen in the results on international average, but all issues have considerably lower means in the Swiss data set. Most social issues are ranked higher compared to environmental issues. For all surveyed environmental and social issues, most companies assessed that stakeholder criticism had not changed during the 2 years preceding the survey (or, for some social issues, that no criticism was expressed) and the fewest companies assessed that it decreased. An increase is attributed to consumer protection by a majority of companies.

12.2.2 Integration

12.2.2.1 Connection to Core Business

Linkages between sustainability and core business are intensive among participating companies. Most Swiss companies state that they connect sustainability to most segments (36.0 %), all segments (28.0 %) or some segments (28.0 %) of their core business, while only 8.0 % declare that they connect sustainability to only a few segments. A similar pattern can be seen on international average, indicating that sustainability is important for most companies participating in the study.

12.2.2.2 Involvement of Organisational Units

In both the Swiss and the international samples CSR/sustainability (including EHS/environment/health/safety) units engage most in sustainability measures (mean: 4.57 and 4.66). In the Swiss data set this is followed by procurement/purchasing (3.96), public relations/corporate communication (3.84) and manufacturing (3.82). The least engagement is attributed to employee council (2.50), financial & management accounting (1.96) and finance (1.95) (the same units as on international average). For these three units, Swiss companies show distinctly less engagement than the international average. Other marked differences between both data sets can be seen for strategic planning, top management and legal/compliance units, with considerably lower values in the Swiss data (see Fig. 12.6).

Swiss companies also assess CSR/sustainability to have the highest promoting impact on sustainability among all organisational units (mean: 4.71), followed

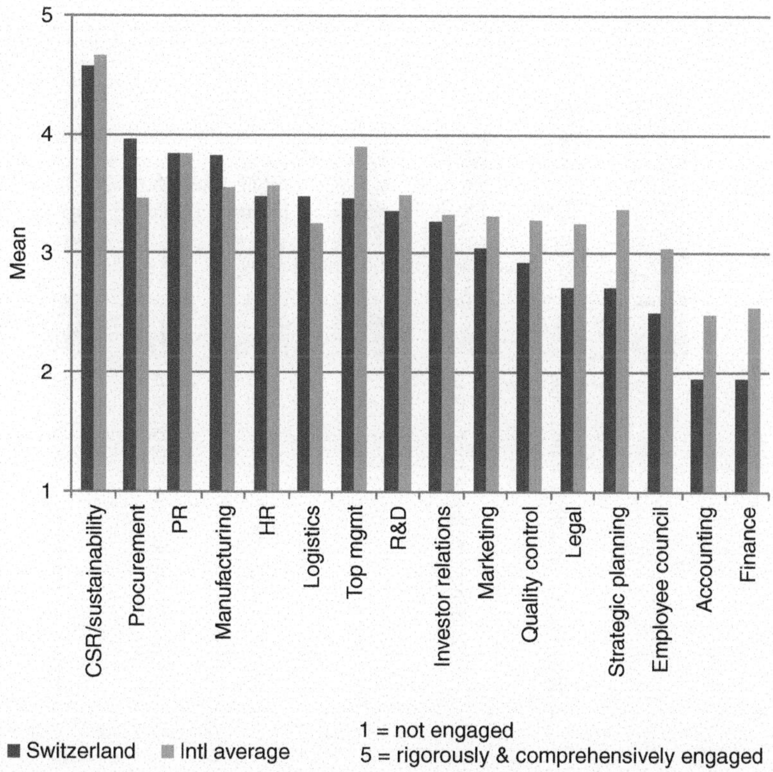

Fig. 12.6 Engagement with sustainability measures of different organisational units (n_{SUI} ranging from 11 to 25, n_{INT} ranging from 286 to 418)

by public relations/corporate communication (4.20) (both of which are similar to the international average) and investor relations (4.00). Financial & management accounting (3.00) and finance (2.84) are the only organisational units which appear to have a relatively indifferent or even inhibiting impact. Compared to the international average, Swiss companies assign a much lower impact to top management and research & development units.

Swiss companies consider CSR/sustainability units (mean: 4.70) most strongly affected by environmental issues, followed by procurement/purchasing (4.08). Finance and financial & management accounting appear least affected. Compared to the international average, impacts on the top management and legal/compliance department are much lower, but most of the other units are also considered as less affected in Switzerland. With regard to social issues, personnel department/HR (4.40) is most strongly affected, followed by CSR/sustainability (4.33) and top management. Finance and financial & management accounting are least affected. Compared to the international average, impacts on strategic planning, finance and

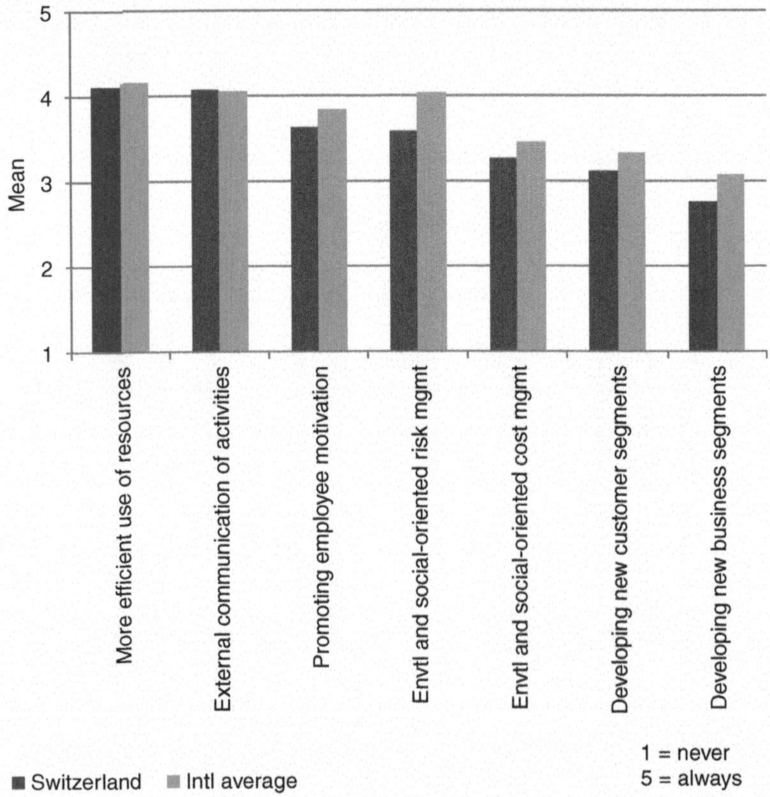

Fig. 12.7 Level of implementation of different sustainability measures (n_{SUI} ranging from 24 to 25, n_{INT} ranging from 397 to 405)

financial & management accounting are considered to be much lower by Swiss companies, but most other units are also assessed to be less affected (the most notable exception being personnel department/HR).

12.2.2.3 Drivers for Sustainability

Producing with more efficient use of resources, e.g. optimising production processes (mean: 4.13) and external communication of environmental and social activities, e.g. sustainability reporting (4.08), were ranked as the most frequently used sustainability measures. Compared to the international average, environmental and social-oriented risk management (e.g. health care at the workplace) is much less often implemented by Swiss companies. Also, the other measures are less frequently implemented (see Fig. 12.7).

12.2.3 Implementation

12.2.3.1 Stakeholder Management

For the management of their stakeholder relations, Swiss and international companies predominantly apply information tools. Of the Swiss companies, 84.0 % stated they inform stakeholders through websites, the press etc. in most cases or in general, while 16.0 % do so on a case-specific basis. Empowerment or delegating decision-making authority is not used by most of the Swiss companies. Only a single company stated that it uses these tools in most cases.

12.2.3.2 Sustainability Management Tools

The level of knowledge of sustainability management tools appears substantial among Swiss companies. A total of 27 of the 79 different sustainability management methods given in the questionnaire are known by 75 % or more of the responding Swiss companies (for an overview of tools see Schaltegger et al. 2007). This is a considerably higher number compared to the international average, where only 15 methods are known by 75 % or more of the companies. Of the methods 61 are known by more than 50 % of the Swiss companies (international average: 64 methods). The most widely known tools are quality management system (QMS), risk analysis, sustainability report, further education, eco-indicators and flexible working time, which are all familiar to more than 90 % of the Swiss companies. Flexible working time, environmental management system and quality management system are also known by 90 % of the companies in the international data set. At the other end of the spectrum, less than one-third of the responding Swiss companies know eco/social ABC analysis, sustainability accounting and socio-eco-efficiency analysis (i.e. the least widely known tools). On international average, only eco-compass was stated to be known by less than one-third of the responding companies.

The sound knowledge of methods is supported by the fact that many tools have been known for a longer period of time. All 31 tools for which this question was asked have been known for more than 2 years by most of the Swiss companies, for many methods this is true for more than 75 % of the companies. Methods that have been known for more than 2 years by a large majority of companies are quality management system (100.0 %), incentive system (95.0 %) and environmental report (94.7 %). The tools known by the fewest companies for more than 2 years are social cost accounting (60 %), social management system (69.2 %) and green supply chain management (69.2 %). Similar trends are seen on international average. On the other hand, the surveyed methods are applied by Swiss companies to a much lower extent. Only seven of the 79 methods are applied by more than 75 % of the responding Swiss companies, 17 methods by more than 50 %. On international average, only four methods are applied by more than 75 % of the companies

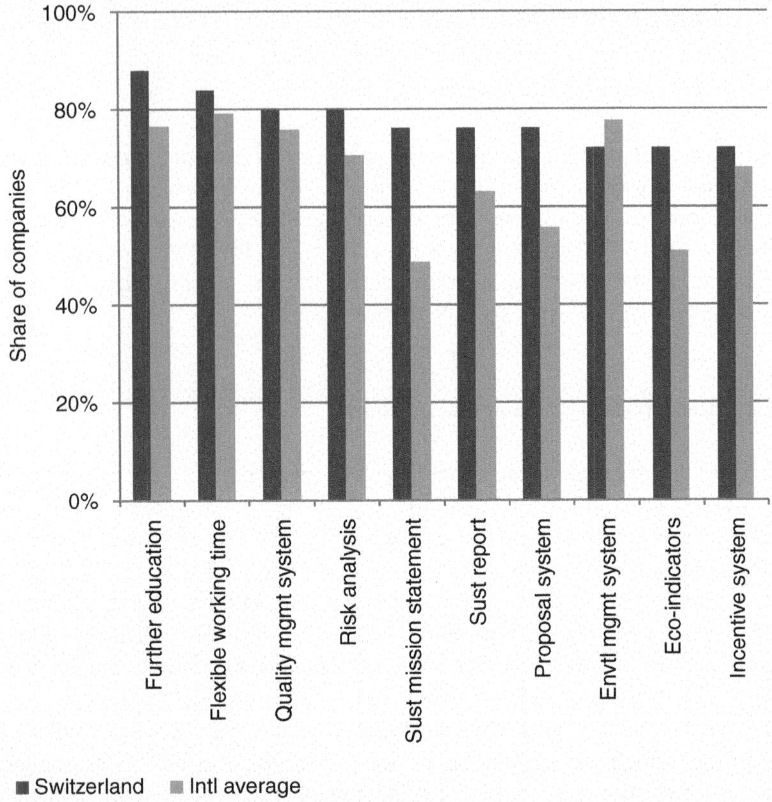

Fig. 12.8 Application of different sustainability management tools (the ten most widely applied tools among Swiss companies are displayed) (per cent; $n_{SUI} = 25$, n_{INT} ranging from 467 to 468)

and 18 by more than 50 %. At least 80 % of the responding Swiss companies apply further education, flexible working time, quality management system and risk analysis, making them the most frequently applied tools (see Fig. 12.8). Less than 5 % of the responding Swiss companies apply social cost accounting, environmental accounting, social accounting, sustainability accounting or opportunity-risk dialogue. Eco/social ABC analysis, socio-eco-efficiency analysis, environmental shareholder value, stakeholder value indicators and eco-compass are not applied by any of the responding Swiss companies. For some tools, marked differences between the Swiss and the international average can be seen. More Swiss companies apply sustainability mission statements (27.4 % above the international average), sustainability management system (+27.2 %) and sustainability control (+22.3 %). Green purchasing is applied by 28.2 % more companies in the international average than in the Swiss data set.

Of the 12 standards and guidelines specified in the questionnaire, Sigma Guidelines (not known by 56.0 %), OHSAS 18001 (48.0 %) and AA 1000 (48.0 %) are still unknown to a large share of Swiss companies. This is similar to the international average, where EFQM (52.1 %) and SA 8000 (48.5 %) are also relatively unknown, whereas OHSAS 18001 is more popular.

In both data sets, most companies do not see a great need for environmental and social management methods to be developed for their organisational units. With regard to the environment, Swiss companies see the greatest need for top management (mean: 3.29), investor relations (3.20) and procurement/purchasing (3.17) and the least need for financial & management accounting (2.36), finance (2.32), and legal/compliance department (2.04). On international average, the greatest need is seen for procurement/purchasing (3.38), research & development (3.19) and logistics/distribution (3.19), the least need for finance (2.69), legal/compliance department (2.66) and financial & management accounting (2.66).

With regard to social management tools, responding Swiss companies see the greatest need for employee council (mean: 3.27), top management (3.23) and personnel department/HR (3.00) and the least need for legal department/compliance (2.10), finance (2.00) and financial & management accounting (1.90).

12.2.3.3 Measurement

Most of the Swiss companies measure their impact on the environmental factors energy consumption (91.7 %), emissions (91.7 %) and water consumption (83.3 %), while only 33.3 % measure their impact on biodiversity. A similar trend can be seen on international average. Also, most of the responding Swiss companies measure their impact on the social factors training and development (95.8 %), workplace/employment (87.5 %) and occupational health and safety (75 %), while only few measure their impact on child labour, forced and compulsory labour, e.g. in the supply chain (43.5 %), freedom of association/right to collective bargaining (31.8 %) and consumer protection (22.7 %). On international average, more than 50 % of the responding companies measure their impact on freedom of association/right to collective bargaining and almost 50 % on consumer protection. Responding Swiss companies state they have improved their relative impact on energy consumption (68.8 %) and occupational safety and health (63.3 %). Few companies state they have worsened their impact on any of the specified environmental and social aspects, but 31.1 % state they have worsened their impact on emissions/wastewater/waste and 30.8 % on transport.

Many Swiss companies measure attractiveness as employer/job satisfaction (68.2 %), risks with relevance for company success (54.5 %), and reputation/image/brand value (54.5 %) with regard to their success or competitive advantage. Innovation (of products, processes etc.) and business model innovation are measured by only a few companies. This is similar to the international sample, where reputation/image/brand value, attractiveness as employer/job satisfaction as well as costs rank highest. For each of the eight issues, the companies measuring these issues were asked to provide the result of that measurement expressed as increase, decrease or no change in competitiveness. For most of the specified issues, the responding companies measuring that issue register no change in competiveness. An increase in competitiveness is registered by the highest percentage of the responding companies for reputation, costs and efficiency. Only few companies register a decrease in competiveness for any of the given issues. On international average, most companies register an increase in competiveness for all of the specified issues.

For Swiss companies both environmental and social management have a relatively positive impact on company success. Success due to environmental management was ranked between 3 and 5, success due to social management between 2 and 5 (where 1 means negative impact and 5 means positive impact). On average, however, Swiss companies rank social management slightly more positive (mean 3.70) than environmental management (3.64). The trend on international average appears similar. The international sample assesses environmental and social management to have a relatively positive impact on company success (the mean for environmental management is 4.00, for social management it is 3.96).

12.3 Conclusion

Compared to the international average, the Swiss data set is to a greater extent dominated by financial and service institutions (while industrial companies are hardly represented), smaller companies in terms of employees, companies with a high share of non-domestic sales and companies which are not family-run. As financial institutions have different needs and perspectives towards some sustainability issues, including applied methods and focus (Pisano et al. 2012), than for example industrial companies, this might help to explain some differences in the results. In this context, the large extent to which sustainability and core business are connected in the participating companies is important, indicating that these companies mainly manage sustainability issues which are material for them. In addition, the smaller average size of the companies in the Swiss data set might have contributed to the specific national characteristics of the results (see Gallo and Christensen 2011).

Also, only companies from the German-speaking part of Switzerland were included, while companies from French and Italian-speaking parts might have other needs and perspectives and as a result might have given different responses to the questions of the study. In the political context of Switzerland, there are marked differences in voter behaviour, indicating that stakeholders might create a different context for companies in the different parts of the country. On the other hand, considering the experience from a 10-year project on sustainability reporting (see e.g., Daub 2007, 2010; Schmidt and Daub 2011), only few companies in the French and Italian-speaking part of Switzerland are engaged in communicating their sustainability performance and so might still be less engaged in sustainability management in general.

Compared to the international average, the Swiss data set reveals that companies perceive stakeholders to have a more inhibiting influence. For example, Swiss companies assigned a lower impact to NGOs, media, the community, consumer organisations and national authorities. It should be stated at this point that in the Swiss data set, the mean average for many responses is lower compared to the international average. This might indicate a general restraint on the part of the responding companies (or even the responding individuals) to assess the questions too positively, i.e. allot higher values. Alternatively, the more inhibiting influence might be due to the high representation of financial companies, which, at least before the financial crisis, have had a different experience with stakeholders than other companies (Furrer et al. 2012). In addition, companies stated criticism from stakeholders was unchanged during the previous years. Still, stakeholder management is dominated by relatively passive tools such as observing and informing stakeholders. Although dialogue instruments are also used, there still seems to be little motivation to empower stakeholders or to include them in decision-making processes, indicating a cautious stance towards stakeholder influence.

Of all the factors surveyed, commitment by top management has the highest promoting impact on the implementation of sustainability in Swiss companies, although their actual impact on, and the degree they are affected by, sustainability issues is relatively limited. Companies do not see a great need to develop environmental and social management methods for their organisational units, but it is notable that the greatest need is seen for top management. Government incentives have the lowest promoting and the highest inhibiting impact, indicating a relatively liberal attitude or the belief that such methods are exhausted and positive initiatives for sustainability need to come from companies or civil society. This is in line with the finding that a lack of personnel capacities and of government incentives has the highest impact on the implementation of sustainability in a company, lack of self-commitment by business the lowest.

Swiss companies expect the most important positive impacts of implementing sustainability to be enhancing and safeguarding corporate reputation as well as employee motivation. In contrast to that, producing with more efficient use of resources and external communication of environmental and social activities are the most frequently used measures that can be seen as drivers for sustainability. Compared to the international average, environmental and social-oriented risk management (e.g. health care at the workplace) is much less often implemented in Switzerland.

The study shows a substantial knowledge of sustainability management tools among Swiss companies (see also ISDC 2012), which indicates that the responsible managers are well educated and have sound competencies. This is supported by the high importance that is attributed to know-how for the implementation of sustainability. However, knowledge could be improved for some accounting and efficiency analysis tools which link sustainability and financial aspects. On the other hand, not many of the tools that are known are actually applied in sustainability management in the companies – and those tools that are applied tend to be general or traditional ones which are also connected to other management aspects (such as quality management, risk analysis, further education, flexible working time, reporting or labels). Compared to the international average, some measurement-related tools like indicators are more widely used in Swiss companies.

References

Berger V, Winistörfer H, Weissert S, Heim E, Schüz M (2012) Swiss Corporate Sustainability Survey 2012. Nachhaltigkeit in Schweizer Unternehmen. Zürcher Hochschule für angewandte Wissenschaften (ZHAW), Winterthur

Daub C-H (2007) Assessing the quality of sustainability reporting: an alternative methodological approach. J Clean Prod 15(1):75–85

Daub C-H (2010) Öffentliche Legitimation durch glaubwürdige Berichterstattung. Ökologisches Wirtschaften 1:30–34

Dun & Bradstreet (Schweiz) AG (2011) Die grössten Industrie-, Handels- und Dienstleistungsunternehmen in der Schweiz 2011. Handelszeitung

Furrer B, Hamprecht J, Hoffmann VH (2012) Much ado about nothing? How banks respond to climate change. Bus Soc 51(1):62–88

Gallo PJ, Christensen LJ (2011) Firm size matters: an empirical investigation of organizational size and ownership on sustainability-related behaviors. Bus Soc 50(2):315–349

ISDC (Interdepartmental Sustainable Development Committee) (2012) Sustainable development in Switzerland – a guide. ISDC, Berne

Pisano U, Martinuzzi A, Bruckner B (2012) The financial sector and sustainable development: logics, principles and actors. Quarterly Report, no 27. European Sustainable Development Network (ESDN), Vienna

Schaltegger S, Herzig C, Kleiber O, Klinke T, Müller J (2007) Nachhaltigkeitsmanagement in Unternehmen: von der Idee zur Praxis: Managementansätze zur Umsetzung von Corporate Social Responsibility und Corporate Sustainability. Centre for Sustainability Management, Lüneburg

Schaltegger S, Harms D, Hörisch J, Windolph SE, Burritt R, Carter A, Truran S, Crutzen N, Ben Rhouma A, Csutora M, Tabi A, Kokubu K, Kitada H, Haider MB, Kim J-D, Lee K-H, Moneva JM, Ortas E, Álvarez-Etxeberria I, Daub C-H, Schmidt J, Herzig C, Morelli J (2013) International corporate sustainability barometer: a comparative study of 11 countries. Centre for Sustainability Management, Lüneburg

Schmidt JEU, Daub C-H (2011) Ein Trend in Wandel. Bedeutung der Nachhaltigkeitsberichterstattung für Unternehmen. Umwelt Perspektiven 6:11–13

Wagner A, Wenk C (2013) Agency versus hold-up: on the impact of binding say-on-pay on shareholder value. Working paper, no 679. National Centre of Competence in Research, Financial Valuation and Risk Management, Zürich

World Bank (2006) Where is the wealth of nations? Measuring capital for the 21st century. The World Bank, Washington, DC

Chapter 13
Managing Responsible and Sustainable Business in the UK

Biswaraj Ghosh and Christian Herzig

Abstract Our study reveals both notable differences and similarities between the average international and UK samples for the issues we investigate in the International Corporate Sustainability Barometer (ICSB). Commitment from top management, corporate philosophy and self-commitment seem to drive corporate sustainability in the UK more strongly than internationally. We also find business case expectations from sustainability adoption to be higher in the UK than internationally. Conversely, the role of legislation and government incentives for implementing sustainability is rated higher in the international sample than by UK companies. Our study also highlights several key points asserted by both the international and the UK samples. These include, firstly, companies' emphasis on managing core environmental and social issues such as the efficient use of energy and water, emissions control and workplace health and safety. However, both cohorts also stress the need for greater contribution towards social issues management. Secondly, whilst organisational units such as CSR/sustainability, top management and public relations are seen to influence responsible and sustainable practices positively, finance and accounting departments have least influence on their implementation.

13.1 Introduction

13.1.1 The UK Context

The United Kingdom (UK) is often seen to be a leading country in socially responsible business (e.g. Bertelsmann Stiftung and GTZ 2007) with an explicit model

B. Ghosh • C. Herzig (✉)
Nottingham Business School, Nottingham Trent University, Nottingham, UK
e-mail: biswaraj.ghosh2013@my.ntu.ac.uk; christian.herzig@ntu.ac.uk

S. Schaltegger et al. (eds.), *Corporate Sustainability in International Comparison*, Eco-Efficiency in Industry and Science 31, DOI 10.1007/978-3-319-06227-3_13,
© Springer International Publishing Switzerland 2014

of business-society relations where businesses undertake a wider role and assume responsibility in societal issues as a consequence of changes in the UK national business systems (including deregulation, privatisation of essential services, socially responsible criteria to access capital and increased public expectations from corporations for social welfare) (Moon 2005; Matten and Moon 2008). Corporate responsibility is reported to have grown and become more institutionalised within the UK, emerging from involvement in more basic forms of responsible business around issues of community to a broader take-up of integrated themes including the environment, workplace, production and products as well as supply chains (e.g. Moon 2005; Herzig and Moon 2011, 2013). Another illustration of the on-going institutionalisation of responsible and sustainable business in the UK is the emergence of business initiatives and associations such as Business in the Community (BITC) with its well-established framework covering Community, Marketplace, Environment and Workplace. Several UK-based multinationals (and several other hundreds of companies) have signed up to what is probably known as the most prominent business association in the area of responsible and sustainable business in the UK. According to one of BITC's most recent reports, sustainable goods and services in the UK are nowadays nearing £200 billion worth (BITC 2013). It is interesting to note that even during the recent economic downturn, UK businesses are seen to have gained from going green and investing in sustainability innovation, for instance clean production technologies, the latter registering a remarkable growth of over 24 % since the financial crisis reached its peak in 2008 (Balch 2013). More and more UK businesses also appear to be investigating innovative ways of achieving resource efficiency for long term business continuity, for example by tapping into cleaner production technologies such as reverse osmosis for long-term water supplies and renewable energy sources (Balch 2013). Our task will be to explore the data gathered within the International Corporate Sustainability Barometer (ICSB) in order to compare and contrast responsible and sustainable business practices in the UK and internationally. In the next section, we introduce the UK sample before presenting our findings.

13.1.2 The UK Sample

The UK sample comprises 36 companies which represent approximately 8 % of all companies participating in the International Corporate Sustainability Barometer (ICSB) survey, coordinated by the Centre for Sustainability Management in Lüneburg, Germany (see Schaltegger et al. 2013). The FAME database was used to obtain information required for the UK study (Bureau van Dijk 2013). The top 250 companies from the UK (ultimate owners with a minimum control path of 50.01 %) were identified on the basis of revenues earned. Out of these, 31 companies stated that they would not participate in any survey and were excluded from the sample, yielding a 16.4 % response rate.

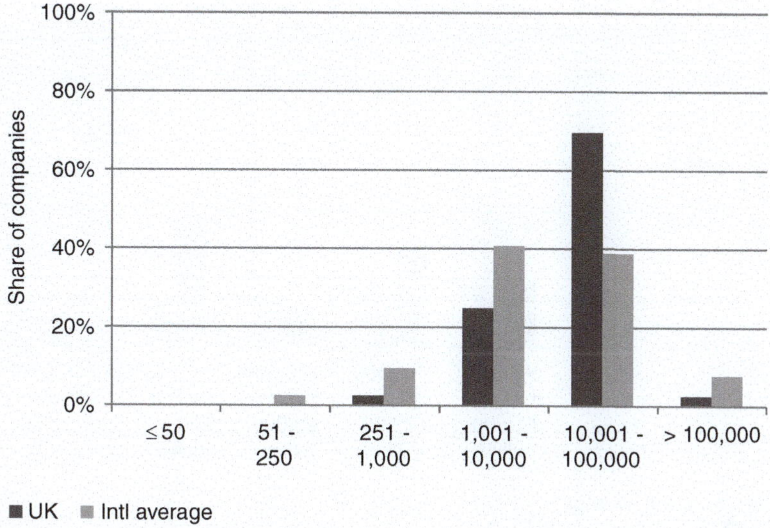

Fig. 13.1 Firm size (number of employees) of the UK sample and the international sample (per cent; $n_{UK} = 36$, n_{INT} ranging from 465)

On the basis of number of employees, the UK sample is characterised by a set of larger sized companies relative to the international sample (see Fig. 13.1). About 72 % of the UK companies employed more than 10,000 employees as opposed to about 46 % for the international sample. However, the UK sample exhibits a slightly lower proportion (2.8 %) of super-sized companies (employing more than 100,000 employees) compared with the international counterpart (7.5 %). The companies in the international group more frequently employed a lower number of employees than their UK counterparts. Around 41 % of the international sample employed between 1,001 and 10,000 people (compared to 25 % for the UK sample) and approximately 10 % of the international sample falls within the 251–1,000 employee category (in contrast to about 3 % for the UK sample). None of the companies in the UK group included in the study employed less than 250 personnel.

Similar to the international average, more than one-third of the UK companies in the sample reported revenues higher than €5,000 million (in the UK 38.9 % vs. 41.2 % for the international group; see Fig. 13.2). Whilst the revenue for each of these companies was between €5,000 and €50,000 million in the UK, about 10 % of the international sample reported revenues higher than €50,000 million.

Almost 31 % of the UK sample reported revenues of between €1,500 and €2,500 million compared with 13.2 % for the international sample. In contrast, none of the UK companies' revenue was below €500 million whereas 13.9 % of the international group generated revenues between €50 and €500 million. In essence, the findings indicate middle to high revenue earning companies within the UK sample in contrast to the international cohort where companies were more evenly distributed across revenue categories.

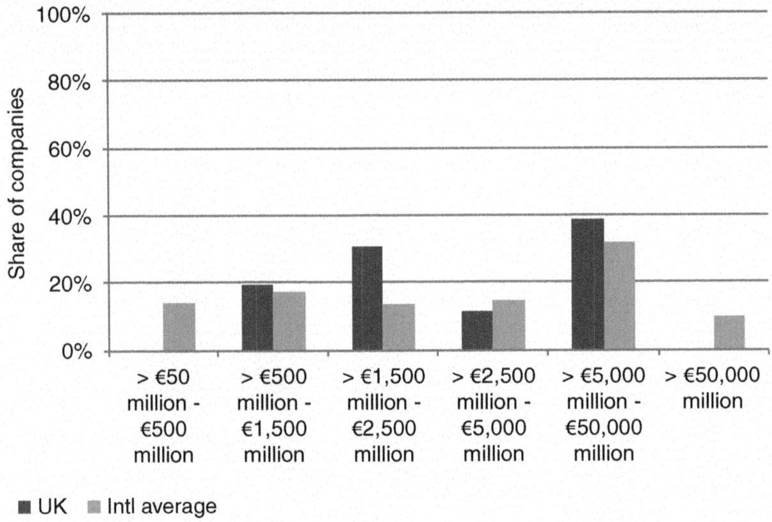

Fig. 13.2 Firm performance (in revenue; per cent of companies; $n_{UK} = 36$, $n_{INT} = 468$)

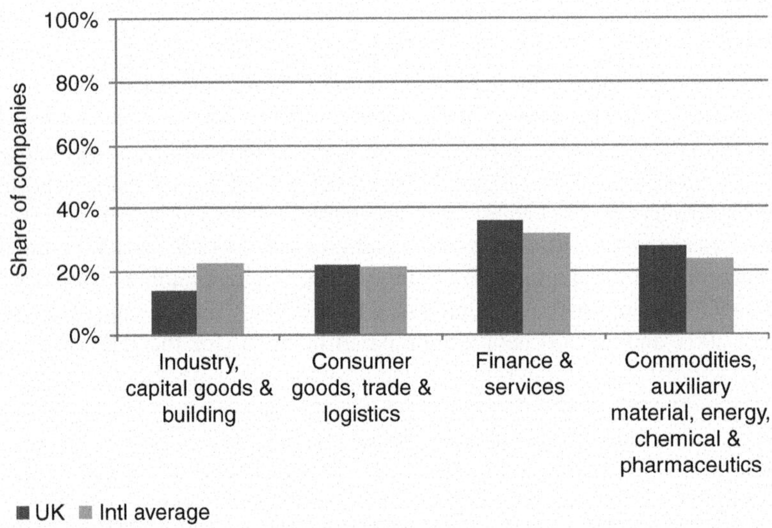

Fig. 13.3 Industry affiliation of companies in the samples (per cent; $n_{UK} = 36$, $n_{INT} = 468$)

In common with the international sample (32.1 %), the largest group of UK companies (over 35 %) belongs to the finance and services sector (see Fig. 13.3). This makes finance and service companies the single largest group in the UK sample which is consistent with statistical findings for the UK economy (ONS 2011). Companies belonging to commodities, auxiliary material, energy, chemical and pharmaceutical industry take the second position for the UK and the international sample. The major difference between the two samples is that businesses belonging to industry, capital goods and building represent nearly 23 % of all companies in the international sample (ranking third), whereas in the UK they account for about 14 % (lowest ranked).

In the next sections, we examine the survey data and present our country-specific findings in comparison with the international group.

13.2 Findings

We present and compare our UK findings with the average results of the international group in relation to the three central focus areas of the ICSB. Firstly, we elaborate on the motives of and drivers for engaging with sustainability (*intention*). We examine the influence internal and external stakeholders have on corporate sustainability and the issues managed by the companies. Secondly, we explore to what extent sustainability is related to the business of the company and the company's prioritisation of stakeholder demands regarding environmental and social issues (*integration*). Finally, we present our findings on the implementation of sustainability through management methods and tools, standards and guidelines as well as the management of stakeholder relations, and we explore which organisational units support sustainability (*implementation*).

13.2.1 Intention

13.2.1.1 Influence of Organisational Units

In relation to their international counterparts, there is a close resemblance in the way the UK cohort rates the influence of organisational units on the implementation of sustainability within their organisations (see Fig. 13.4). Both the UK and the international samples have rated their CSR/sustainability, top management as well as the public relations/communications units as having the strongest influence on sustainability implementation. On the other hand, finance and accounting units are perceived by both groups as having the least influence on promoting sustainability implementation, with UK ratings being even slightly lower than the international average. However, neither at the UK nor at international level were these units regarded to inhibit the implementation of sustainability.

13.2.1.2 Influence of External Stakeholders

In contrast to the fairly consistent influence of various internal stakeholder groups, there are some differences in the way the UK group rates the influence of their external stakeholders on the implementation of sustainability when compared to

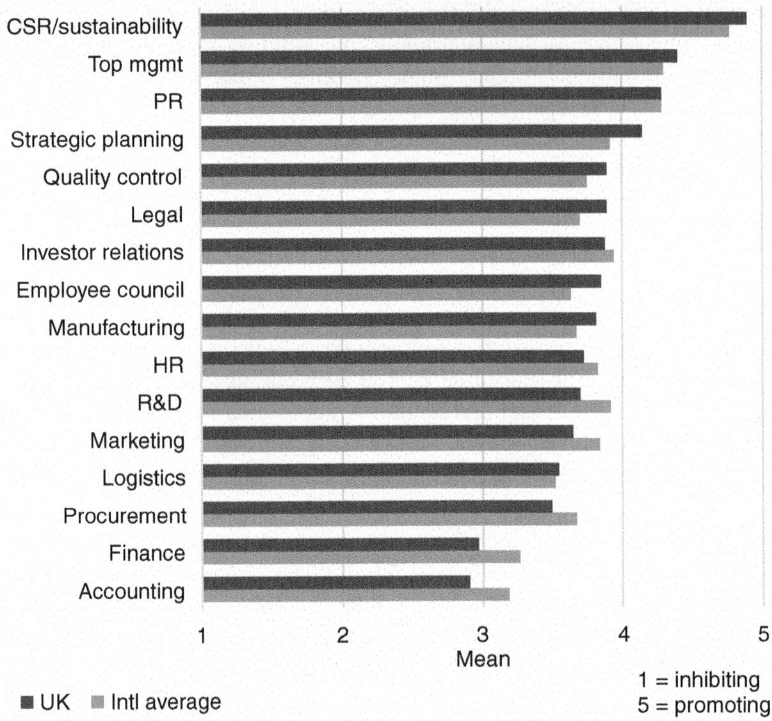

Fig. 13.4 Influence of organisational units (n_{UK} ranging from 16 to 36, n_{INT} ranging from 325 to 460)

the international ratings (see Fig. 13.5). Whilst NGOs, environmental and social organisations are regarded to be most influential by both cohorts, the international sample reports that the media/public have the second highest impact on corporate sustainability, followed by international authorities and communities. The UK group, however, reported that communities have a far greater impact on their corporate sustainability management than the media/public or international authorities. In the UK, communities were a close second in the ranking of the role of external stakeholders for corporate sustainability. As the next most influential stakeholder groups, the media/public, national authorities/legislators, consumer organisations and competitors were ranked ahead of international authorities. Trade unions, suppliers, banks and insurance companies have been rated least influential in promoting sustainability at corporate level by both the UK and international samples.

13.2.1.3 Drivers for and Barriers to Sustainability

Commitment by top management and corporate philosophy are regarded as the most important drivers for sustainability, with even stronger emphasis given to these

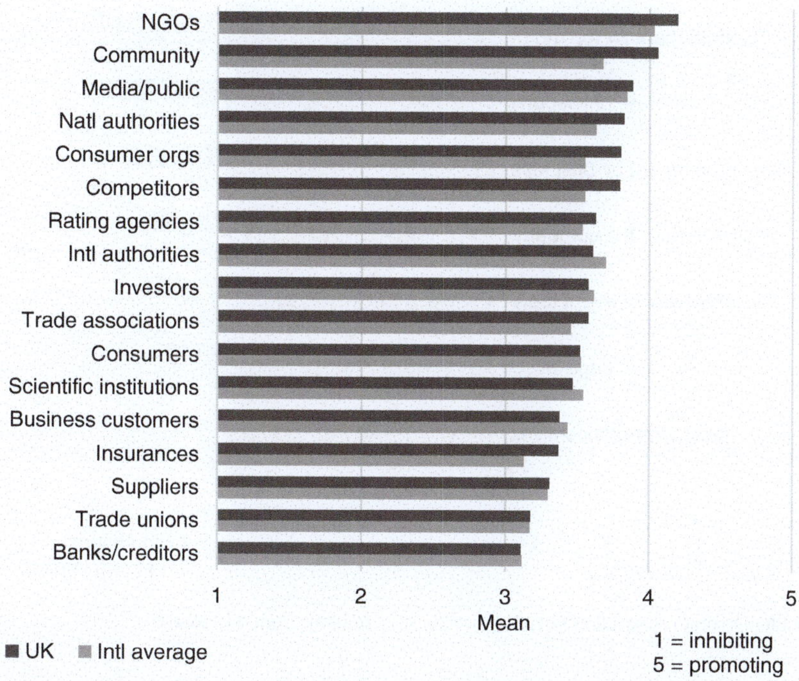

Fig. 13.5 Influence of external stakeholders (n_{UK} ranging from 28 to 36, n_{INT} ranging from 393 to 450)

factors by the UK sample than by their international counterparts (see Fig. 13.6). Other important factors supporting the implementation of sustainability include the practicability of sustainability management methods and a company's know-how in this area. One important difference was that the UK group gave the least importance to government incentives and legislative pressure, while self-commitment to sustainability issues was perceived as the third most important driver for implementing sustainability. In contrast, the role of legislation was identified as being third most important by the international group, whilst self-commitment was ranked amongst the least influential factors. Government incentives were nominated to have the least impact for both groups.

Lack of financial resources was identified as having the strongest negative impact on sustainability within UK companies (without illustration). The consequence of a double-dip recession that was feared to hit the UK during the economic downturn might explain this. Unlike the UK group, the international group ranked a lack of personnel capacity as the major barrier to the adoption of sustainability. Nevertheless, financial incapacity was nominated by both cohorts as a major barrier to sustainability implementation.

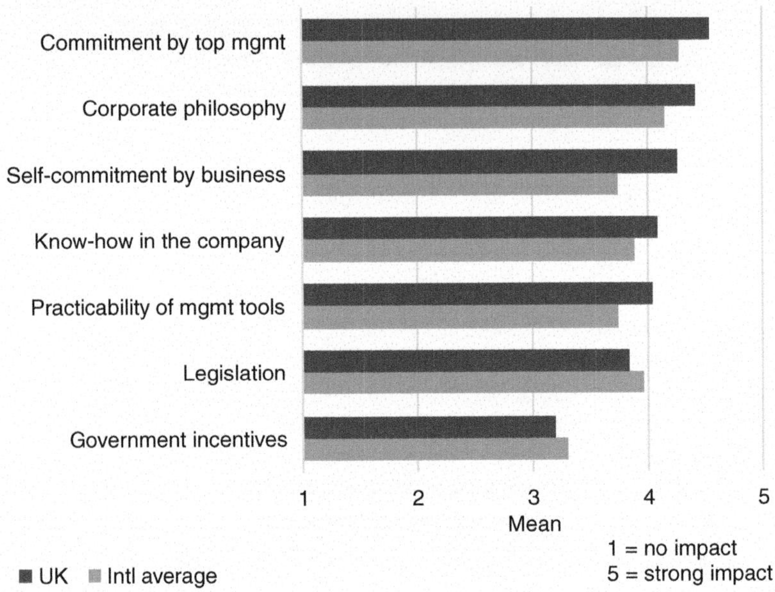

Fig. 13.6 Drivers for sustainability ($n_{UK} = 36$, n_{INT} ranging from 446 to 460)

Table 13.1 Environmental issues ($n_{UK} = 36$, n_{INT} ranging from 455 to 463)

	UK			International		
	Management of issue		Stakeholder demands	Management of issue		Stakeholder demands
Environmental issue	Average	Rank	Rank	Average	Rank	Rank
Energy consumption	4.81	1	2	4.57	1	1
Emissions/wastewater/ waste	4.61	2	1	4.29	2	2
Water consumption	4.17	3	4	3.88	4	4
Material consumption	4.06	4	3	3.95	3	3
Transport	3.89	5	5	3.66	5	5
Biodiversity	2.83	6	6	2.86	6	6

13.2.1.4 Issues

Overall, the UK group appears to manage environmental and social issues more closely than their international counterparts. However, the relative importance of environmental and social issues is fairly consistent among the two groups (see Tables 13.1, 13.2 and 13.3).

The environmental issues most closely managed are energy consumption and emissions, wastewater and waste, followed by water consumption in the case of the UK (ranked third) while ranking fourth internationally. The impact of (environmentally-oriented) NGOs is one possible explanation why companies at

Table 13.2 'Internal' social issues ($n_{UK} = 36$, n_{INT} ranging from 459 to 461)

	UK			International		
	Management of issue		Stakeholder demands	Management of issue		Stakeholder demands
'Internal' social issue	Average	Rank	Rank	Average	Rank	Rank
Health and safety	4.72	1	1	4.61	1	1
Workplace/employment	4.64	2	2	4.44	3	2
Training/development	4.61	3	4	4.47	2	4
Diversity/equal opportunity	4.39	4	3	4.16	4	3

Table 13.3 'External' social issues ($n_{UK} = 36$, n_{INT} ranging from 442 to 449)

	UK			International		
	Management of issue		Stakeholder demands	Management of issue		Stakeholder demands
'External' social issue	Average	Rank	Rank	Average	Rank	Rank
Consumer protection	3.94	1	1	3.73	2	1
Freedom of association/ right to collective bargaining	3.89	2	2	3.80	1	2
Child, forced and compulsory labour	3.75	3	3	3.49	3	3

national and international levels manage these issues closely. NGOs were earlier reported to be the most dominant stakeholder group influencing the implementation of sustainability. However, this seems not to apply to the topic of biodiversity. Stakeholder pressure for managing biodiversity is reported to be the lowest and as a single issue was the environmental issue nominated to be least closely managed in both groups of companies.

Turning to the management of social issues, there is again consistency between the two groups in that both manage 'internal' social issues (i.e. more traditional and employee-related issues such as occupational health and safety, workplace/employment, diversity as well as training and development, average >4.35) more closely than 'external' social issues (i.e. broader, in part more recent social issues such as consumer protection and child labour, average <4.00).

In the UK, occupational health and safety, workplace/employment as well as training and development are reported to be among the three most important social issues, followed by diversity and equal opportunity, consumer protection, and freedom of association/the right to collective bargaining. Child labour, forced and compulsory labour receive least importance in both samples.

Finally, the influence of external stakeholders on the approach to sustainability was assessed in terms of stakeholder demands to manage environmental and social issues (see Tables 13.1, 13.2 and 13.3). Generally – and perhaps unsurprisingly – great similarities exist between perceived stakeholder pressures to

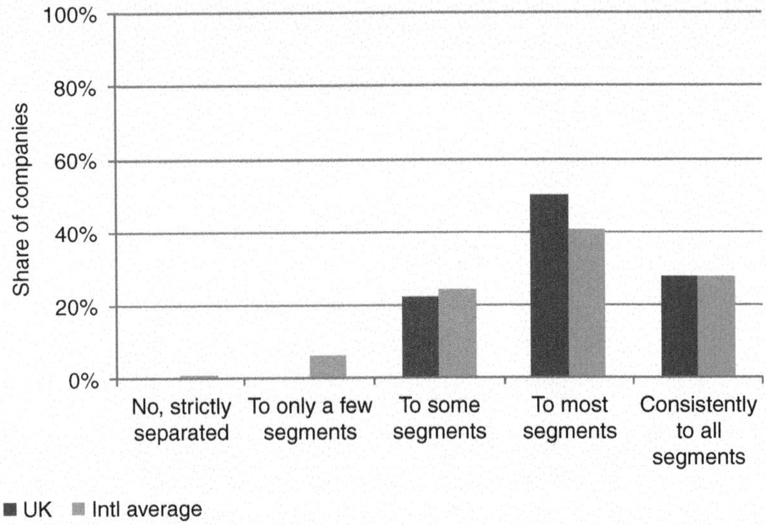

Fig. 13.7 Integration of sustainability within core business segments (per cent; $n_{UK} = 36$, $n_{INT} = 457$)

manage environmental and social issues and the extent to which these are actually managed by the companies at both UK and international levels.

13.2.2 Integration

13.2.2.1 Integration into Core Business

All of the UK companies in the sample reported to have at least some segments of their core business connected to sustainability, unlike a few companies in the international sample which stated that sustainability and core business were separated or only to a small extent connected within their organisation (see Fig. 13.7). More than three-quarters of the UK cohort (77.8 %) linked their sustainability commitment to most or all segments of their core business. This is higher than the international average (66.9 %), although a similar percentage of companies from the UK and the international samples claimed to consistently link sustainability to all core business segments.

13.2.2.2 Business Cases for Sustainability

Overall, business case expectations are higher in the UK than internationally (see Fig. 13.8). Most prevalent was the expectation that sustainability would enhance and safeguard reputation, motivate employees, foster innovation and improve internal

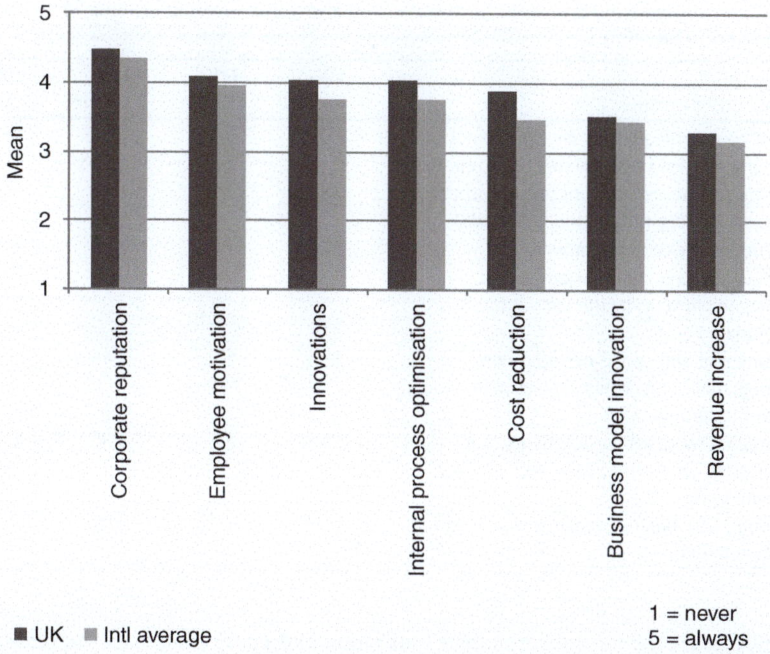

Fig. 13.8 Business cases expectations (n_{UK} ranging from 34 to 36, n_{INT} ranging from 450 to 458)

processes. The UK group also emphasised more strongly the role of sustainability for cost reduction effects. Revenue generation and business model innovation were among the least expected effects of engaging in sustainability for both cohorts.

Slight variations are observed in the ways certain sustainability measures are integrated. For the UK, external communication of non-financial performance (4.17) takes precedence while internationally the importance of external communication of non-financial performance (4.06) is second to efficiency measures. Developing new business segments related to sustainability and developing an environmentally and socially-conscious customer base receive the least importance both in the UK and internationally (see Table 13.4).

13.2.2.3 Addressing Environmental and Social Issues Raised by Stakeholders

The implementation of sustainability can be analysed according to issues managed by companies in response to stakeholder concerns. As presented earlier, stakeholder concerns for the significant environmental issues included energy consumption, emissions, water usage and material consumption. Over one-third of the respondents felt that stakeholder criticism for the aforementioned issues have remained unchanged over the past 2 years, both in the UK and internationally (see Fig. 13.9).

Table 13.4 Sustainability measures (1 = never, 5 = always) (n_{UK} ranging from 35 to 36, n_{INT} ranging from 397 to 405)

Sustainability measures	UK average	UK ranking	Intl average	Intl ranking
External communication of environmental and social activities (e.g. sustainability reporting)	4.17	1	4.06	2
Producing with more efficient use of resources (e.g. optimising production processes)	4.11	2	4.16	1
Promoting employee motivation	4.03	3	3.85	4
Environmental and social-oriented risk management (e.g. health care at the workplace)	3.86	4	4.03	3
Environmental and social-oriented cost management (e.g. using cost-effective recycling products)	3.54	5	3.46	5
Developing new customer segments (e.g. promoting environmentally and socially oriented products)	3.44	6	3.35	6
Developing new business segments related to sustainability	2.94	7	3.09	7

However, stakeholder concerns over emissions and energy use have increased the most relative to the other environmental issues investigated over the past 2 years, applicable for both cohorts. A small percentage of the international sample claimed reductions in criticism over the past 2 years for each of the environmental issues investigated whilst a small percentage of UK respondents claimed a reduction in criticism only for issues relating to transport.

Turning to the social issues investigated, about two-fifths of the respondents reported that stakeholder criticism has remained unchanged for most of the issues, both in the UK and internationally (see Fig. 13.10). At least one-fifth of the companies in both samples were of the view that criticism for social issues including employment, health and safety, diversity and consumer protection has increased over the 2 years preceding the survey. A decrease in criticism of social issues is generally reported by fewer companies in the UK compared with companies in the international sample.

13.2.3 Implementation

13.2.3.1 Managing Stakeholder Relations

Our findings indicate minimal direct participation of stakeholders in corporate advisory boards or decision-making processes in general, which is true for

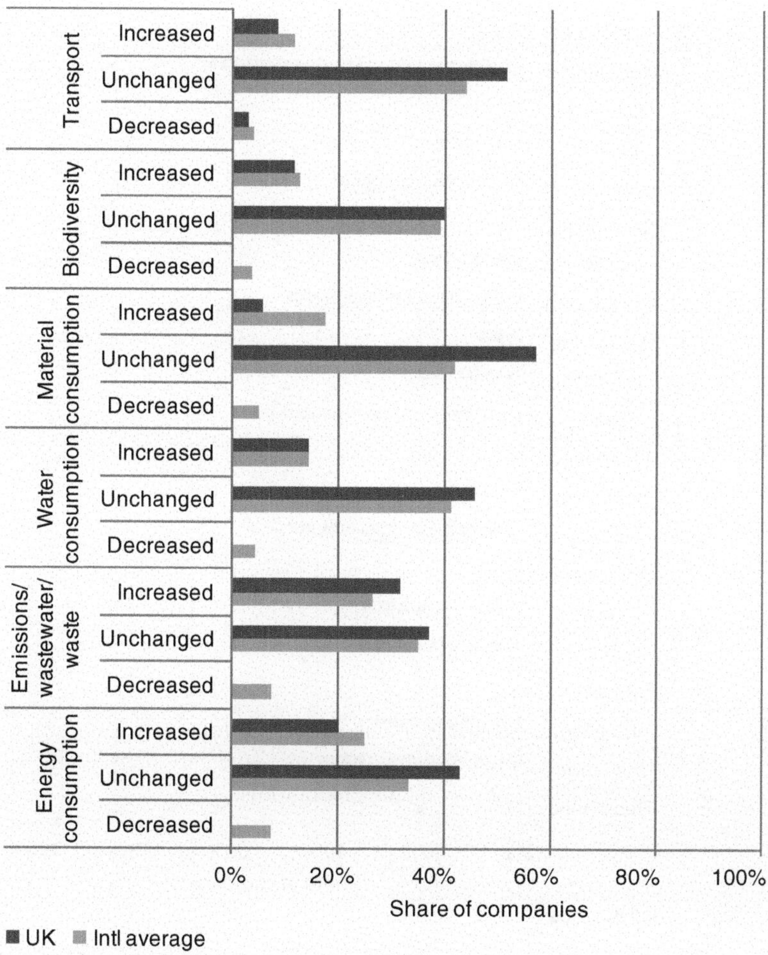

Fig. 13.9 Stakeholder criticism for environmental issues over the past 2 years (per cent; $n_{UK} = 35$, n_{INT} ranging from 447 to 453)

both cohorts. Stakeholder management usually comprises informing stakeholders through company websites and press, taking higher precedence than stakeholder observation or dialogue. Stakeholder empowerment and engagement in decision-making receive the lowest average means of stakeholder management. Direct stakeholder participation involving cooperation and participation in advisory boards is however prominent both in the UK and international samples on a case-to-case basis but not used generally (see Table 13.5).

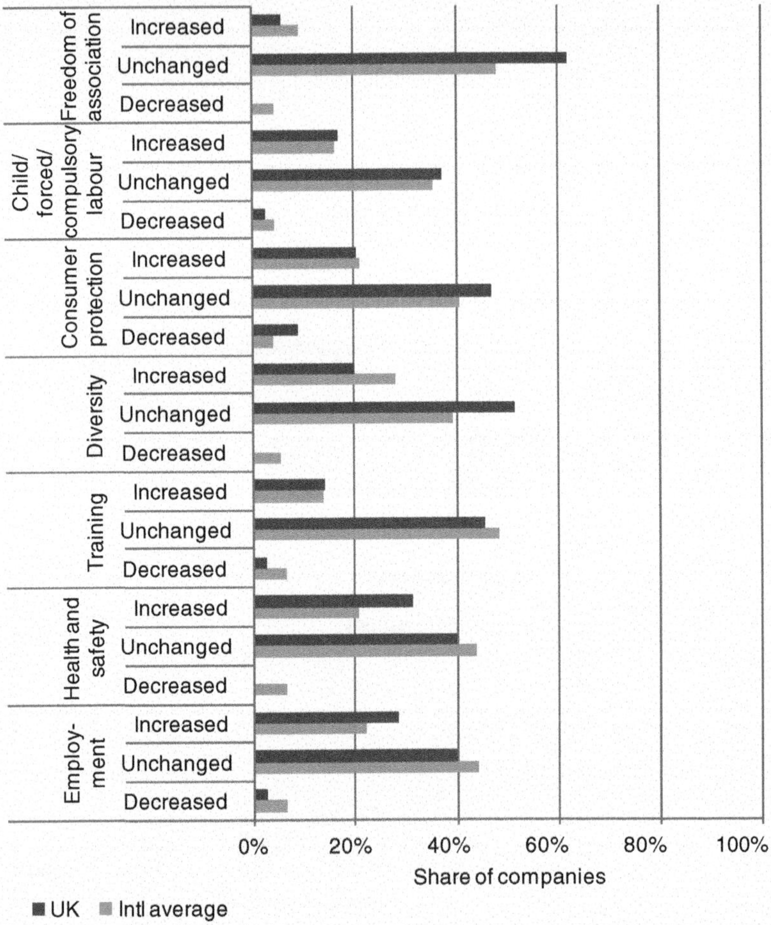

Fig. 13.10 Stakeholder criticism for social issues over the past 2 years (per cent; n_{UK} ranging from 34 to 35, n_{INT} ranging from 436 to 451)

13.2.3.2 Methods of Sustainability Management

There is considerable similarity in terms of the type of tools which are seen to be of higher importance to both groups of companies (see Table 13.6). Frequent use of management methods is reported by both cohorts with regard to corporate community involvement, employees, management systems, external reporting and mission statements. However, differences do exist in the degrees of their application. For example, application of methods to manage corporate community involvement (e.g. corporate giving) is much higher in the UK than internationally.

Table 13.5 Managing stakeholder relations (n_{UK} ranging from 35 to 36, n_{INT} ranging from 438 to 458)

	UK (%)			Internationally (%)		
	Not used	Used on a case-specific basis	Used in most cases or in general	Not used	Used on a case-specific basis	Used in most cases or in general
Observing stakeholders	8.3	38.9	52.8	8.1	48.4	43.5
Informing stakeholders	0.0	16.7	83.3	1.5	21.4	77.1
Dialogue with stakeholders/seeking advice	8.6	45.7	45.7	10.1	53.8	36.0
Involvement, consideration in decision-making process	30.5	50.0	19.4	27.5	53.4	19.1
Cooperating, networking to develop joint solutions	8.3	66.7	25.0	16.1	61.2	22.7
Empowerment	44.4	41.7	13.9	43.0	49.5	7.4
Delegating decision-making authority	54.3	34.3	11.4	62.6	32.9	4.6

Table 13.6 Application of selected methods of sustainability management

Type of tool	Tool	UK %	Intl %	UK/Intl	Diff
Management system	Environmental information system			UK	★
	Environmental management system			UK	★
	Social management system			UK	★
	Quality management system			UK	☆
	Sustainability management system			UK	☆
Control	Eco-control			Intl	★
	Human resource control			UK	☆
	Sustainability control			UK	☆
	Sustainability balanced scorecard			UK	☆
Audit	Eco-audit			Intl	★
	Social audit			UK	★
	Sustainability audit			UK	★
Design	Eco-design/design for environment			UK	★
	Sustainable design			UK	★
Risk management	Early detection			UK	☆
	Risk analysis			UK	★
	Scenario analysis			UK	★
	Cross-impact-analysis			UK	☆
Supply chain management/ purchasing	Green purchasing			UK	☆
	Green supply chain management			UK	★
	Sustainable supply chain management			UK	★
Accounting	Material flow analysis			UK	★
	Product line analysis			UK	☆
	Eco-efficiency analysis			UK	★
	Eco-balance/life-cycle assessment			Intl	☆
	Product carbon footprint			UK	☆
	(Eco/social) ABC analysis			UK	☆
	Socio-eco-efficiency-analysis			UK	☆

(continued)

Table 13.6 (continued)

	Material and energy flow accounting			UK	
	Material flow cost accounting			Intl	
	Environmental cost accounting			Intl	
	Social cost accounting			Intl	
	Eco-budgeting			UK	
	Eco-investment accounting			Intl	
	Environmental accounting			UK	
	Social accounting			UK	
	Sustainability accounting			UK	
Indicator	Eco-indicators			Intl	
	Eco-efficiency indicators			UK	
	Social indicators			UK	
	Socio-efficiency indicators			UK	
	Sustainability indicators			UK	
	Environmental shareholder value			UK	
	Stakeholder value indicators			UK	
Benchmark-ing	Eco-benchmarking			UK	
	Social benchmarking			UK	
	Sustainability benchmarking			UK	
Mission statement	Environmental mission statement			UK	
	Social mission statement			UK	
	Sustainability mission statement			UK	
Report	Environmental report			UK	
	Social report			UK	
	Human resource report			UK	
	Sustainability report			UK	
Marketing	Eco-marketing			Intl	
	Social marketing			UK	
	Sustainability marketing			UK	
Label	Eco-label			Intl	

(continued)

Table 13.6 (continued)

	Social/fair label	1–25 %	1–25 %	UK	☆
	Sustainability label	1–25 %	1–25 %	UK	☆
Sponsoring	Eco-sponsoring	1–25 %	1–25 %	Intl	☆
	Social/cultural sponsoring	26–50 %	51–75 %	Intl	☆
	Sustainability sponsoring	26–50 %	26–50 %	UK	☆
Employees	Incentive system	76–100 %	51–75 %	UK	⯪
	Flexible working time	76–100 %	76–100 %	UK	☆
	Proposal system	26–50 %	51–75 %	Intl	⯪
	Further education	76–100 %	76–100 %	UK	☆
Corporate community involvement	Corporate giving	76–100 %	51–75 %	UK	★
	Corporate/employee volunteering	76–100 %	51–75 %	UK	⯪
	Community advisory panel	26–50 %	1–25 %	UK	☆
	Corporate citizenship	51–75 %	26–50 %	UK	⯪
Stakeholder management	Stakeholder dialogue/forum	51–75 %	26–50 %	UK	☆
	Opportunity-risk dialogue	26–50 %	1–25 %	UK	⯪

Legend

Application 1–25 %
Application 26–50 %
Application 51–75 %
Application 76–100 %

UK: Difference larger in the UK
Intl: Difference larger internationally

★ Difference >25 %
⯪ Difference >10 % ≤ 25 %
☆ Difference ≤10 %

13.2.3.3 Standards and Guidelines

We find similarities in the type of standards and guidelines known and adopted by businesses internationally and in the UK (see Table 13.7). The ISO 14001 standard is well known and implemented by the highest percentage of companies in both samples, followed by ISO 9000. It is interesting to note that the new ISO 26000 norm on CSR is known by far more companies in the UK, although on average applied by a much lower number of companies than internationally. With regard to guidelines, GRI is the most widely applied guidance document, much ahead of the UN Global Compact guidelines, both internationally and in the UK.

Table 13.7 Standards and guidelines for sustainability management

Standards and guidelines	UK		Intl	
	Only known	Known and applied	Only known	Known and applied
ISO 14001	16.7	80.6	16.7	73.1
ISO 9000	25	69.4	21.4	66.7
ISO 26000	77.8	8.3	58.5	15.4
EMAS	66.7	11.1	48.9	17.1
OECD guidelines	55.6	16.7	49.8	18.4
GRI	30.6	61.1	27.4	52.8
UN Global Compact	52.8	36.1	40.6	34.4
EFQM (incl. S-EFQM)	44.4	16.7	36.8	11.1
AA 1000	58.3	22.2	37.4	12.2
OHSAS 18001/BS 8800	38.9	50.0	29.5	34.6
SA 8000	61.1	11.1	41.7	9.8
Sigma guidelines	52.8	19.4	30.3	16.0

Legend

Application 1–25 %

Application 26–50 %

Application 51–75 %

Application 76–100 %

13.2.3.4 Involvement of Organisational Units

In order to manage various environmental and social issues, the participating companies were asked which organisational units were involved in the implementation of sustainability (without illustration). CSR/sustainability units, followed by top management and public relations/communications units, are most comprehensively engaged in sustainability measures. Accounting and finance units have the least involvement in sustainability management. These observations apply to companies from both the UK and worldwide, although we find top management in UK companies more rigorously engaged than in the international cohort. Overall, the results seem to mirror our observation on the most influential organisational units impacting on sustainability (as illustrated earlier).

Companies were also asked for which organisational units they would see a need for development of suitable environmental and social management methods (see Table 13.8). To address environmental issues more effectively, we find both

Table 13.8 Demand for new environmental management methods in organisational units

	UK ranking			Intl ranking	
Organisational unit	Need for new envtl methods	Impacted by envtl issues	Organisational unit	Need for new envtl methods	Impacted by envtl issues
Procurement/purchasing	1	8	Procurement/purchasing	1	4
Logistics/distribution	2	3	Logistics/distribution	3	7
Research and development	3	6	Research and development	2	3
Strategic planning	4	5	Strategic planning	4	8
Top management	5	4	Manufacturing	5	2

international and UK companies accentuating the strongest need for new management methods with regard to their purchasing and procurement units. Both cohorts of participants also reported that research and development, logistics and strategic planning units were in need of new environmental management methods.

Both UK and international respondents reported that the top five organisational units which require new social management methods were procurement/purchasing, top management, personnel department/HR, strategic planning and CSR/sustainability (see Table 13.9). However, to manage social concerns more effectively through new management tools, UK respondents pointed towards their procurement and purchasing units first, whilst, in contrast, international participants ranked this unit fifth. The latter cohort asserted HR/personnel department as the foremost, while UK respondents ranked it third.

Overall, we observe that both groups of companies stress the need for new methods for similar organisational units although there are differences in the way these units are seen to be impacted by environmental and social issues (see Tables 13.8 and 13.9). Whilst in some cases the demand for new management methods can be deduced from the high level of perceived importance of these units (strong environmental and/or social impact; e.g. logistics/distribution, top management), differences between the two rankings in other instances might result from anticipation of their future importance. For example, despite the current relatively moderate level to which procurement/purchasing is seen to be affected by environmental and social issues, there is a clear interest in new methods to perhaps manage these issues more effectively in the future. Finance and accounting units, which are ranked lowest with regard to the extent to which organisational units are impacted by environmental and/or social issues (without illustration), also rank very low regarding the demand for new management methods however. This, together with the low take-up of existing methods in this area (see Table 13.6), is surprising given that these units provide relevant information for organisational decision-making processes.

13.3 Conclusion

We elaborate on three main conclusions drawn from our comparative analysis.

- Responsible and sustainable business as an integrative and overlapping management concept in the UK

The UK group reports wide engagement in responsible and sustainable business activities. Various environmental and social issues are cited to be closely managed and most companies in the sample referred to management systems and other methods within the areas of community involvement, employees, mission statements, reporting and risk management through which these issues are managed. Whilst there is large similarity to the international sample in this regard, there is stronger emphasis on concepts and labels often used to describe the more traditional account of CSR by the UK sample than by the international sample. This is for example

Table 13.9 Demand for new social management methods in different organisational units

	UK ranking			Intl ranking	
Organisational unit	Need for new soc methods	Impacted by soc issues	Organisational unit	Need for new soc methods	Impacted by soc issues
Procurement/purchasing	1	10	Procurement/purchasing	5	11
Top management	2	2	Top management	2	2
Personnel department/HR	3	3	Personnel department/HR	1	3
Strategic planning	4	6	Strategic planning	4	6
CSR/sustainability	5	1	CSR/sustainability	3	1

reflected in the wider involvement in corporate giving and employee volunteering. The UK group also refers more frequently to the concept of corporate citizenship and rates corporate philanthropy to be a stronger driver for engagement than the international sample. However, the UK group does not only buy into basic levels of community involvement and philanthropy, which largely reflected CSR's reputation in the early 1980s; we also find 50 % of UK companies in the sample linking sustainability to most of the core business segments, which is higher than the international average. Business expectations from engaging with sustainability are manifold and overall also higher in the UK than internationally. Company benefits in the form of intangible assets such as reputation and employee motivation were most frequently cited by companies, followed by improvement of internal processes and innovation. In a similar vein, and perhaps reflecting the more explicit nature of responsible business in the UK, respondents see legislative pressure and government incentives as amongst the least influential drivers, while regulatory pressure and guidelines are ranked third internationally. However, for UK companies in the sample self-commitment played a bigger role than for the international sample.

All of these results seem to suggest that managing responsible and sustainable business in the UK can be described as an integrative and overlapping management practice, i.e. it entails notions of various concepts of corporate responsibility and sustainability and seems to be well embedded in the organisations surveyed.

- External stakeholders are genuinely important drivers for corporate sustainability, but questions remain about opportunities for stakeholder engagement

Responding to external stakeholder groups is seen to play a pivotal role in corporate sustainability initiatives by companies from both samples. NGOs, environmental and social organisations are found to have the strongest influence on businesses. Given that corporate community involvement has been a long-standing feature of UK business in the last two or three decades, it is perhaps less surprising that community was ranked in second place when UK respondents were asked which stakeholders primarily drive their sustainability activities. Within the international sample, companies ranked the media/public second and community only fourth. Notably, more companies in the UK than internationally also report the use of management methods of community involvement (e.g. corporate giving, employee volunteering).

Whilst the companies' view of the importance of stakeholders could suggest that stakeholders lie at the heart of the sustainability management process, a closer look at the applied methods raises questions about the deeper meaning and truthfulness of companies' engagement with their stakeholders. Stakeholder empowerment and engagement in decision-making processes receive least attention within stakeholder management. Instead of stakeholder dialogue, management seems to prefer informing stakeholders through websites or the press. If direct forms of stakeholder participation exist, they are used on a case-by-case basis rather than on a regular basis, both in the UK and internationally.

- Future challenges for management centre on the accounting and finance function, supply chain issues and involvement of strategic planning

In both groups of respondents, similar organisational units are perceived to have the highest promoting influence for sustainability: CSR/sustainability, top management and public relations/communications. The least influence is by far reported from accounting and finance. Based on this, is there an inherent danger in the evident underestimation of the role of accounting and finance since these organisational units and functions are the backbone of every organisation and provide relevant information for much decision-making?

From the survey responses of both samples it also emerged that the management of sustainability issues within the production/manufacturing process appears to be well understood and supported through methods described. Moreover, the management of issues related to the supply chain is regarded as the area in which there is one of the highest needs for new environmental and social management methods. Both groups of companies agreed in that there was a need for new tools for procurement/purchasing (environmental and social issues) and logistics and distribution (environmental issues). Finally, we note a high interest in methodical support for top management and strategic planning to enhance the management of environmental and social issues and support strategic renewal and implementation processes.

References

Balch O (2013) Sustainable innovation could be worth £100bn for UK companies. The Guardian, 4 Jul 2013. http://www.theguardian.com/sustainable-business/sustainable-innovation-worth-100-billion-companies. Accessed 5 Sept 2013

Bertelsmann Stiftung, GTZ (2007) The CSR navigator. Public policies in Africa, the Americas, Asia and Europe. Bertelsmann Stiftung/GTZ, Gütersloh

Bureau van Dijk (2013) Fame. Detailed information on UK and Irish companies. https://fame.bvdinfo.com/version-20131011/Home.serv?product=fameneo. Accessed 19 Feb 2012

Business in the Community (BITC) (2013) Fortune favours the brave. BITC, London

Herzig C, Moon J (2011) Corporate social responsibility, the financial sector and economic recession. Financial Services Research Forum, Nottingham

Herzig C, Moon J (2013) Discourses on corporate social ir/responsibility in the financial sector. J Bus Res 66:1870–1880

Matten D, Moon J (2008) "Implicit" and "explicit" CSR: a conceptual framework for a comparative understanding of corporate social responsibility. Acad Manag Rev 33(2):404–424

Moon J (2005) United Kingdom. An explicit model of business-society relations. In: Habisch A, Jonker J, Wegner M, Schmidpeter R (eds) Corporate social responsibility in Europe. Discovering national perspectives of corporate citizenship. Springer, Berlin/Heidelberg, pp 51–65

ONS (2011) UK business: activity, size and location 2011. http://www.ons.gov.uk/ons/dcp171780_236119.pdf. Accessed 1 Jul 2013

Schaltegger S, Harms D, Hörisch J, Windolph SE, Burritt R, Carter A, Truran S, Crutzen N, Ben Rhouma A, Csutora M, Tabi A, Kokubu K, Kitada H, Haider MB, Kim J-D, Lee K-H, Moneva JM, Ortas E, Álvarez-Etxeberria I, Daub C-H, Schmidt J, Herzig C, Morelli J (2013) International corporate sustainability barometer: a comparative study of 11 countries. Centre for Sustainability Management, Lüneburg

Chapter 14
The Case of Corporate Sustainability in the United States of America

John Morelli and Dorli Harms

Abstract Being the third largest country by population and the largest economy today, the United States of America (US) impacts changes and developments on a global scale while US companies also face national and international challenges such as resource consumption, emissions control, as well as health and safety concerns. The findings on the current state and progress of sustainability management and Corporate Social Responsibility (CSR) in the US reveal that social issues are closely managed and that the stakeholders demand according engagement by the companies. The US companies surveyed that are engaged in sustainability are also quite society-oriented. Striving for legitimacy, risk-orientation and compliance therefore can be seen as predominant motives for dealing with sustainability issues. At the same time the companies have the potential to seize market opportunities by being more innovative with respect to their business models, products and services.

14.1 Introduction

14.1.1 The US American Context

The United States of America (US) is the largest economy (World Bank 2013) and the third largest country by population (US Census Bureau 2013) worldwide, so it is not surprising that it plays a considerable role in international trade, setting

J. Morelli (✉)
Department of Civil Engineering Technology, Environmental Management & Safety, Rochester Institute of Technology, Rochester, NY, USA
e-mail: john.morelli@rit.edu

D. Harms
Centre for Sustainability Management (CSM), Leuphana University, Lüneburg, Germany
e-mail: dharms@uni.leuphana.de

S. Schaltegger et al. (eds.), *Corporate Sustainability in International Comparison*, Eco-Efficiency in Industry and Science 31, DOI 10.1007/978-3-319-06227-3_14,
© Springer International Publishing Switzerland 2014

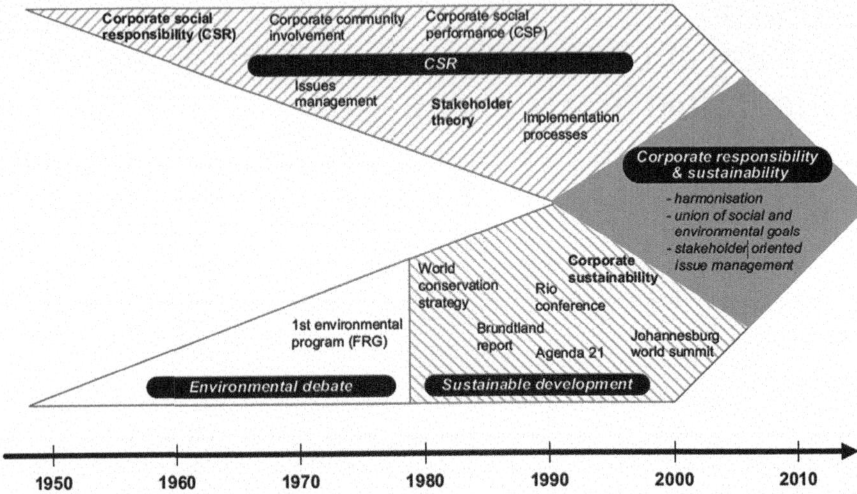

Fig. 14.1 Historical development of the debate on CSR and sustainable development (Hansen 2010: 8; based on Loew et al. 2004)

of global trends and in its consumption of the world's resources. At the same time US companies are still challenged by limited resources, and with concerns regarding the community and employee safety and well-being. Taking responsibility for staff and society are cornerstones of the nowadays internationally-commonly used term 'corporate social responsibility' (CSR) which has its origin in the US during the 1950s when Bowen (1953) published his book 'Social Responsibilities of the Businessman' (Loew et al. 2004; Crane et al. 2014). Parallel to this stream, as it is illustrated in Fig. 14.1, the deliberation on sustainable development evolved, having its roots in the environmental debate. Moreover, in the 2000s the term of corporate sustainability has become more common (Loew et al. 2004).

Today, terms such as corporate citizenship, CSR and corporate sustainability are quite intertwined and sometimes used as synonyms although they have different emphases: while CSR refers to responsibilities of a company, for instance, corporate sustainability has a strong link to the Brundtland report (WCED 1987), which implies that economic, environmental and social aspects are supposed to be simultaneously integrated into the company's conventional management. In so doing, the management of corporate sustainability strives for a contribution to the sustainable development of the company as well as of society and the economy (Loew et al. 2004; Schaltegger and Burritt 2005).

14.1.1.1 Voluntary Initiatives

When companies manage corporate sustainability there is also the distinction between mandatory and voluntary initiatives for contributing to sustainable development. Depending on national circumstances such as historic background, laws and people's attitudes, companies from different countries may engage more or less voluntarily in corporate sustainability. In the US, voluntary corporate initiatives, for instance, seem to be common (Hess et al. 2002; Kruger 2005). Another aspect to mention is that the US is very large and does not have as centralised a structure as compared to other countries. Therefore it is not easy to find one-fits-all solutions to societal or political challenges, which in turn supports the fact that the economic system builds more on individual and voluntary than mandatory solutions. However, concerning elements of sustainability, this tendency might apply more to environmental initiatives than to occupational health and safety as will be discussed below.

14.1.1.2 Workplace Health and Safety in the US

Today, organised labour in the US represents only 12 % of the non-farm, wage and salary workforce and compared to the typical range of 20–40 % in Europe is among the weakest in this respect of developed democracies (Wright and Rogers 2010). While organised labour in the US historically has been lower than that of many comparable countries, union membership grew in the US from the end of the nineteenth century to reach approximately 36 % of the workforce by 1954 (Blanchflower and Freeman 1990) and along the way played a major role in the workplace safety movement in the US. This was particularly so with regards to the overturn of employer-biased laws which supported doctrines of contributory negligence, which held that an employee who accepts a job also assumes the risks associated with whatever the work might be (Goetsch 2005). Throughout the 1960s, the US saw passage of major health and safety related legislation focusing on a variety of specific industries. By the end of the 1960s, however, the injury and death tolls in the US were still high and rising, largely due to the fact that many industries were still not covered by this flurry of legislation (LaDou 1986). This led to the passage of the Occupational Safety and Health Act of 1970, representing the most comprehensive and significant health and safety legislation. It is uniquely characterised by its General Duty Clause that states in addition to complying with all standards, rules, regulations and orders issued pursuant to the Act, that "each employer shall furnish to each of his employees employment and a place of employment which are free from recognised hazards that are causing or are likely to cause death or serious physical harm to his employees" (29 United States Code § 654, 5(a)1).

There is considerable inconsistency among countries with respect to how they implement their occupational health and safety programmes. While inspections and enforcement are part of all of them, the strength and vigour with which sanctions are imposed vary. While Kendall's (2006) review of health and safety programmes in 5 countries does not indicate which country's programme is most effective at safeguarding worker health and safety, it appears to indicate that the US may be most assertive in imposing fines on employers who violate its standards.

14.1.1.3 Diversity and Assimilation in the US

There are many ways of measuring diversity in a country, including quantifying the variety of national, cultural and ethnic origins of its people. On these bases, the US probably ranks among the most diverse nations globally. Historically, the trend in the US has been to assimilate people from different backgrounds into the US culture, in fact, recent studies demonstrate that immigrants to the US assimilate better than in other countries, including most European countries (Vigdor 2008). One might argue that such assimilation reduced diversity and with respect to other metrics (e.g., language, dress, etc.) it does. However, assimilation in the US has been a two-way street and "changes in the customs and practices of the native-born can promote cultural assimilation just as easily as changes among the foreign-born" (Vigdor 2008: 3).

"Pasta, salsa, sausage, and egg rolls are now as common place on American dinner tables as corn, pumpkin, and turkey. Soccer is now a national pastime, at least among youth, and millions of sports fans cheer the hundreds of immigrants who are members of Major League Baseball" (Myers and Pitkin 2010: 1).

The recent election and then re-election of the nation's first African-American president provides, perhaps, the most obvious recent evidence of this sometimes slow but effective assimilation process.

14.1.2 The US Sample

The research findings on the current state and progress of sustainability management in the US presented here were obtained by means of an online survey among large US companies (by revenue) which was conducted between July and August 2012. The survey was part of the International Corporate Sustainability Barometer (ICSB) coordinated by the CSM in Lüneburg, Germany (see Schaltegger et al. 2013). The US companies of different sectors were identified by making use of the database Fortune 500 list (CNN Money 2012). For the survey, company managers in charge of sustainability respectively environment, health and safety (EHS) or CSR were contacted by email, via the company websites or phone and were asked to fill in the online questionnaire. In total, 179 links to the online survey were sent out, which yielded 34 responses, equalling a response rate of 19 %.

Tables 14.1 and 14.2 illustrate the sample characteristics of the US data in comparison to the international average based on the responses of all 11 countries surveyed for the ICSB.

With regard to the international survey, the US companies rank among the largest since about 67 % generate an annual revenue between 5,000 and 50,000 million euros whereas this is only true for about 31 % of the companies of the international average. A similar picture emerges for the number of employees since about 80 % of the responding companies from the US have 10,001–100,000 employees whereas

Table 14.1 Annual revenue[a] ($n_{US} = 34$, $n_{INT} = 468$)

Annual revenue in Euro	US (%)	Intl average (%)
>50–500 million	0	13.9
>500–1,500 million	0	17.1
>1,500–2,500 million	2.9	13.2
>2,500–5,000 million	23.5	14.5
>5,000–50,000 million	67.6	31.4
>50,000 million	5.9	9.8
Total	*100*	*100*

[a]Figures include total assets/gross premiums for banks/insurance companies

Table 14.2 Number of employees ($n_{US} = 33$, $n_{INT} = 465$)

Number of employees	US (%)	Intl average (%)
≤50	0	0.4
51–250	0	2.8
251–1,000	0	9.7
1,001–10,000	3.0	40.9
10,001–100,000	81.8	38.7
>100,000	15.2	7.5
Total	*100*	*100*

less than 40 % of the companies of the international average have such a large number of employees. The results also show that the surveyed US companies are to a large majority not family-run (93.9 %) while 80.8 % of the international average is not family-run business.

In contrast, the distribution of the following four sectors derived from the company's core business activities appears quite similar between the US data and the international average:

- Industry, capital goods and building
- Consumer goods, trade and logistics
- Finance and services
- Commodities, auxiliary material, energy, chemical and pharmaceutics

Both on the US and on international levels almost all sectors have a share between 21 and 33 % so that the percentage appears quite proportionate; only the US sector 'industry, capital goods & building' takes a smaller share of about 15 %.

14.2 Analysis

In this section the US results are presented and discussed in more detail to depict the current state and progress of sustainability management and CSR.

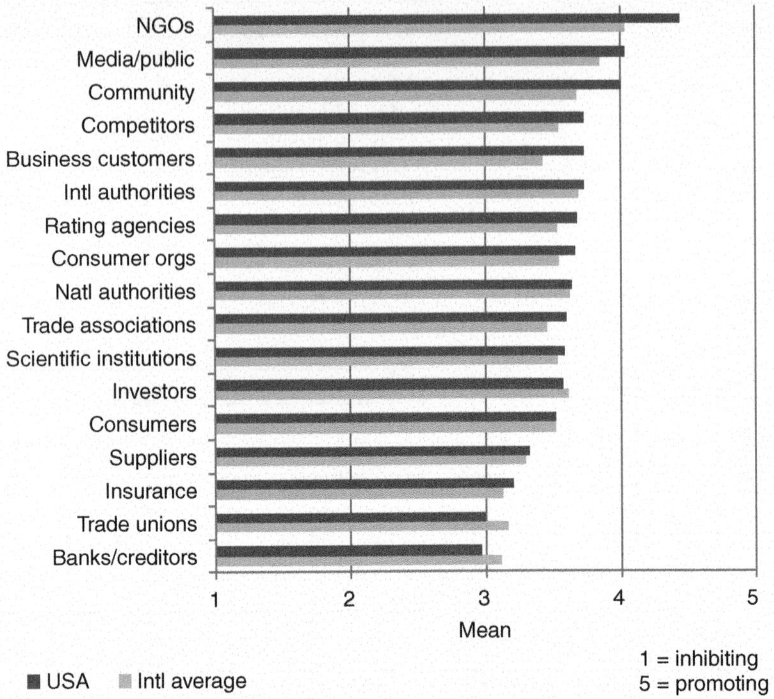

Fig. 14.2 Influence of external stakeholders (n_{USA} ranging from 24 to 34, n_{INT} ranging from 393 to 450)

14.2.1 *Intention*

Depending on the motivation of a company's sustainability commitment, different strategic patterns for dealing with different sustainability issues may be appropriate.

Figure 14.2 illustrates that the US companies assess the society-oriented stakeholders, non-governmental organisations (NGOs), and media/public and community as promoting corporate sustainability the most. The values for vendors/business-to-business customers as well as competitors are also higher than the international average – although on a lower rank than the first three stakeholders. Banks, by contrast, rank slightly lower than the international average. All other stakeholders are viewed similarly by the US companies and their international peers.

The high relevance of striving for societal legitimacy is also demonstrated by the fact that of all expected positive impacts of corporate sustainability surveyed, enhancing and safeguarding *corporate reputation* receives the highest values in the US companies – which is in line with the international average. The other impacts are in order of descending frequency: *employee motivation*, *innovation (products/processes)*, *internal process optimisation*, *cost reduction* and *business model innovation* which are all viewed similarly by the US companies and

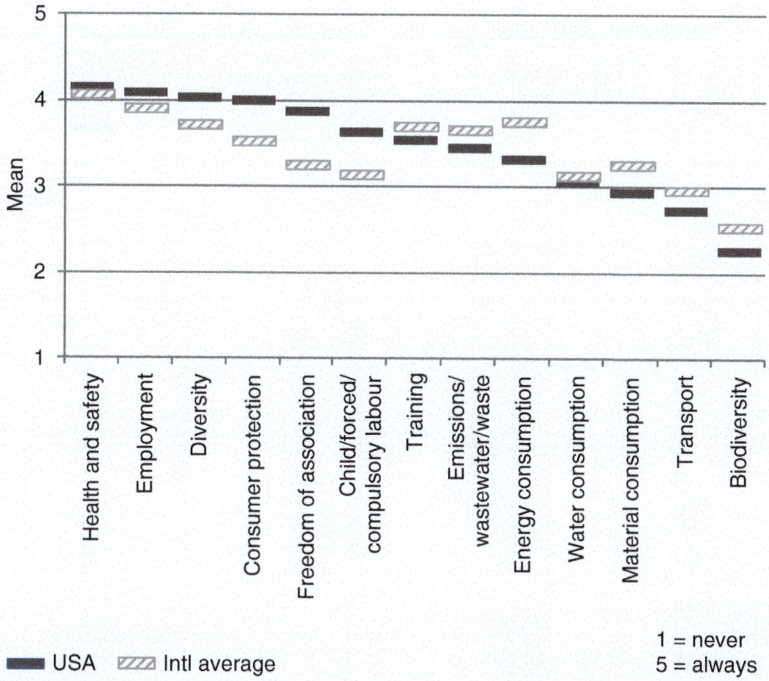

Fig. 14.3 Stakeholder demands to manage sustainability issues (n_{USA} ranging from 32 to 34, n_{INT} ranging from 443 to 461)

consistent with the international average. The potential impact of *revenue increase*, however, receives a lower value in the US companies.

Figure 14.3 shows for which sustainability issues stakeholders demand engagement by companies. Strikingly, almost all social issues which were investigated in this survey rank higher in the US than, on average, in the other countries. Only the social issue *training and development* ranks a bit lower in the US compared to the international average. The US results also reveal that social issues appear to be more often demanded by stakeholders than environmental issues. This outcome might be explained by the fact that stakeholders in the US expect companies to take responsibility for their own employees as well as for society more often than in other countries where more government regulation on CSR initiatives exists.

Apart from the question on stakeholder demands the companies were also asked to evaluate how closely they manage these environmental and social issues (see Fig. 14.4). Here, it becomes obvious that the US results are mostly above the international average, except for the environmental issues *material consumption* and *biodiversity*. In contrast to the findings on stakeholder demands it also becomes clear that the environmental issue *energy consumption* is as extensively managed as are the social issues *occupational health and safety*, *workplace/employment* and *diversity*. The fact that the US companies state they closely manage the environmental issues *energy consumption* and *emissions/wastewater/waste*, although stakeholder

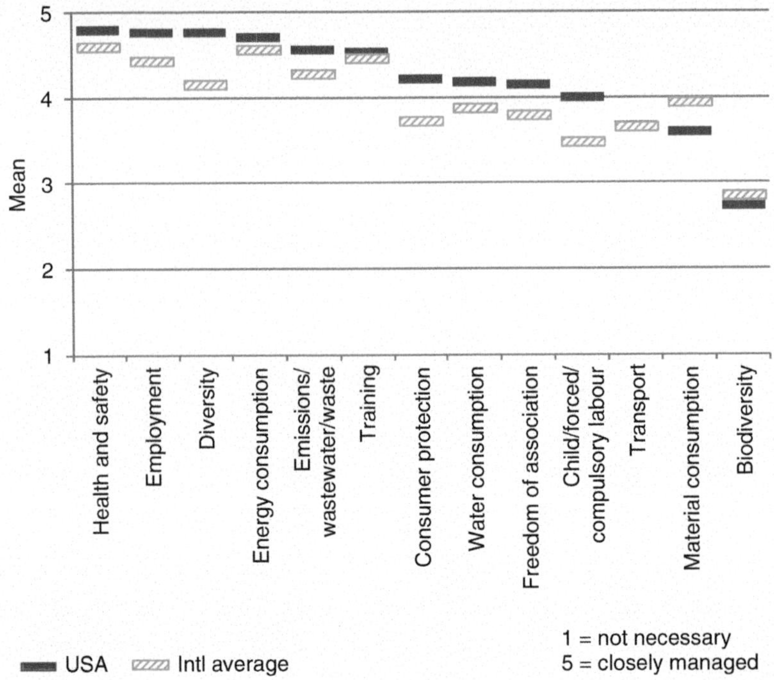

Fig. 14.4 Managed sustainability issues (n_{USA} ranging from 32 to 34, n_{INT} ranging from 442 to 463)

demands for these issues are observed only on a moderate level, indicates that the companies are self-motivated to deal with these issues (e.g. Creyts et al. 2010).

Looking at the sustainability issues that US companies assess as relevant for their business in the next 5-10 years reveals that resource-related issues such as *energy* including renewable energy, (greenhouse gas) *emissions* and *water* play an important role for these companies. In addition, issues related to the *supply chain* as well as *employee well-being*, *diversity* and *transparency* (of reporting, business, workforce) are brought up by several companies. Issues that seem to be of more specific interest because they are mentioned only by one company are, for instance, *soil conservation*, *weather, veterans/disabled veterans* and *indigenous peoples. Bridging technology gap*, *new products* and *ecosystem services valuation* are additional topics companies mention.

14.2.2 Integration

The hallmark of corporate sustainability is how well a company is able to integrate environmental and social policies and programmes into its core business, how

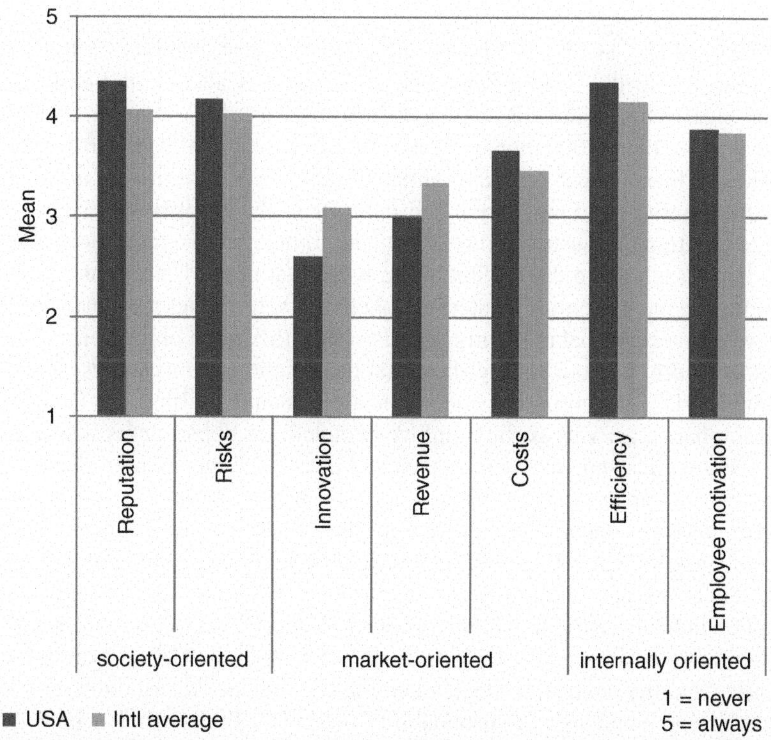

Fig. 14.5 Drivers of business cases for sustainability ($n_{USA} = 33$, n_{INT} ranging from 397 to 405)

well it relates sustainability to its value creation and profit-making activities and if it involves all organisational units in this process to ensure full organisational commitment.

Similar to the results for the international average, most of the US companies claim to link sustainability to most or all segments of their core business (about 67 % of the US companies and 69 % of the companies on international average). Slightly different is only that the US companies link sustainability as often to most segments as they link it to all segments (each about 33 %), whereas on international average linking sustainability to all segments is indicated by just 28 % of the companies while the connection to most segments is declared by 41 %.

By linking sustainability with its core business a company is able to create business cases for sustainability, i.e. generating positive business effects through voluntarily or mainly voluntary outstanding environmental and social performance (Schaltegger 2011; Schaltegger et al. 2012). With regard to different drivers of business cases for sustainability, illustrated in Fig. 14.5 (*reputation*, *risk management*, *innovation*, etc.), it can be noticed that society-oriented and internally oriented drivers are more often addressed than market-oriented drivers – this is true for the companies from the US and also on international average. Figure 14.5 also underlines the relevance of *reputation* for the US companies as well as a risk-orientation of these companies whereas *revenue* and *innovation*, in particular, are considered less as drivers of a business case for sustainability on the national level compared to the international data. This result is also confirmed in a study by MIT Sloan Management Review and Boston Consulting Group in 2012 which examined more than 4,000 managers/executives internationally (Kiron et al. 2013). They showed that North American companies rank lowest in terms of sustainability-focused business model innovation and in terms of the number of business model innovators stating that sustainability adds to the company's profit (Kiron et al. 2013). This might be related to the finding of the ICSB that in the US top management is not among the most promoting internal stakeholders. It implies that sustainability is not viewed as a core goal of the organisation but rather as an obligation.

Further findings of our survey show that the functional units CSR/sustainability (including EHS), PR/corporate communication, manufacturing, the legal department/compliance as well as the employee council are departments which engage expansively with sustainability measures and that the engagement of these top five departments is above international average. On the one hand, it is reasonable that CSR/sustainability departments engage the most with sustainability because it is their main task to deal with environmental and social issues. On the other hand, the finding that corporate communication and the legal department are strongly engaged is another indication for the high relevance of preserving corporate reputation, reducing risks and striving for legitimacy. In contrast, in the US companies finance, accounting, quality control, strategic planning and logistics/distribution rank lowest in terms of engagement with sustainability measures. In particular with regard to strategic planning and quality control these findings differ from the international average where strategic planning holds rank five and quality control ranks in the middle.

14.2.3 Implementation

The implementation of corporate sustainability can be regarded as the extent to which relevant stakeholders are involved in the process, how systematically sustainability management tools are known and applied and how precisely the success of corporate sustainability is measured.

In all countries surveyed the companies frequently inform their stakeholders, whereas intensive forms of stakeholder management such as empowerment and delegating decision-making authority are applied less frequently (see Fig. 14.6: in most cases/in general and Fig. 14.7: case-specific). Yet, country-specific differences exist as the US responses are above average with regard to most forms of stakeholder management. Figure 14.6 reveals, for instance, that the majority of the US companies surveyed observe and inform their stakeholders. Having a dialogue with their stakeholders and cooperating with them is also a commonly used form by more than one third of the companies. The tendency that US companies perform an active stakeholder management is also reflected in the findings illustrated in

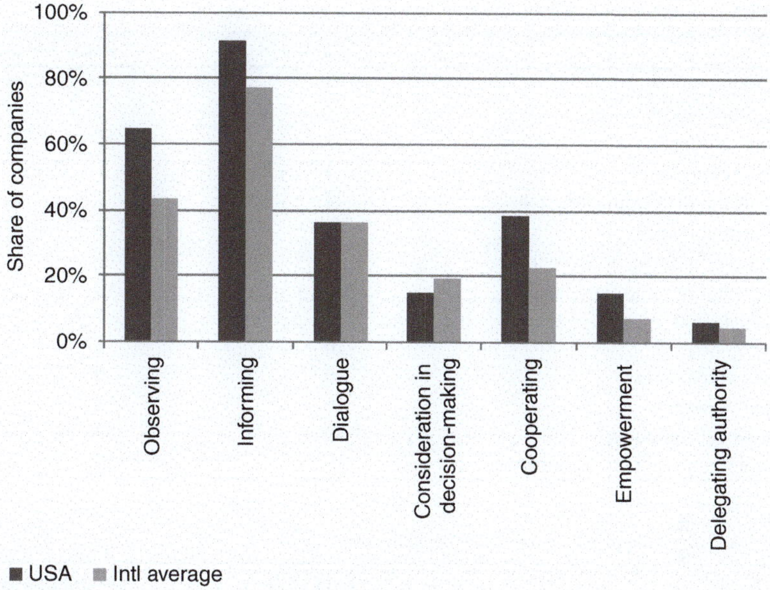

Fig. 14.6 Management of stakeholder relationships: in most cases or in general (per cent; n_{USA} ranging from 33 to 34, n_{INT} ranging from 438 to 458)

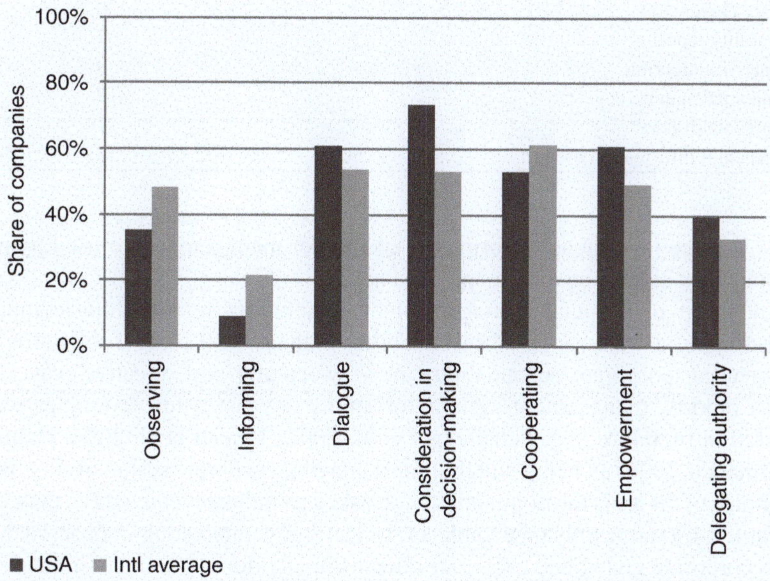

Fig. 14.7 Management of stakeholder relationships on a case-specific basis (per cent; n_{USA} ranging from 33 to 34, n_{INT} ranging from 438 to 458)

Table 14.3 Top ten known sustainability management tools

Tool	Rank: US	Rank: Intl average
Corporate giving	1	16
Sustainability report	2	4
Corporate citizenship	2	21
Flexible working time	4	1
Environmental management system	4	2
Environmental report	4	8
Environmental information system	7	11
Environmental mission statement	7	10
Further education	7	7
Incentive system	10	6
Corporate volunteering	10	9
Green purchasing	10	12
Environmental statement	10	17
Sustainable supply chain management	10	20

Table 14.4 Top ten applied sustainability management tools

Tool	Rank: US	Rank: Intl average
Corporate giving	1	8
Flexible working time	2	1
Environmental management system	3	2
Further education	3	3
Corporate citizenship	3	20
Sustainability report	6	7
Corporate volunteering	6	9
Environmental statement	8	17
Environmental report	9	11
Environmental information system	9	13

Fig. 14.7, since less intensive forms of stakeholder management are used by many US companies at least on a case-specific basis.

In addition to a sound management of stakeholder relationships companies can also apply a wide range of sustainability management tools if they strive for implementing corporate sustainability in a systematic and effective way. Tools such as reports, management systems or audits are just a few examples widely discussed in literature (e.g. Schaltegger et al. 2002; Tencati et al. 2004; Hahn and Scheermesser 2006) as being suitable for managing corporate sustainability. Based on a choice of 79 tools the companies were asked to indicate which tools they know (see Table 14.3) and apply (see Table 14.4). Here, it is interesting to observe that in the US corporate giving and corporate citizenship, i.e. tools which are quite closely connected to CSR (Epstein 1989), rank among the top five whereas these tools hold rank 8 and lower on the international level. In contrast, however, flexible working time is the most applied tool on the international level whereas it holds rank 6 in the US companies.

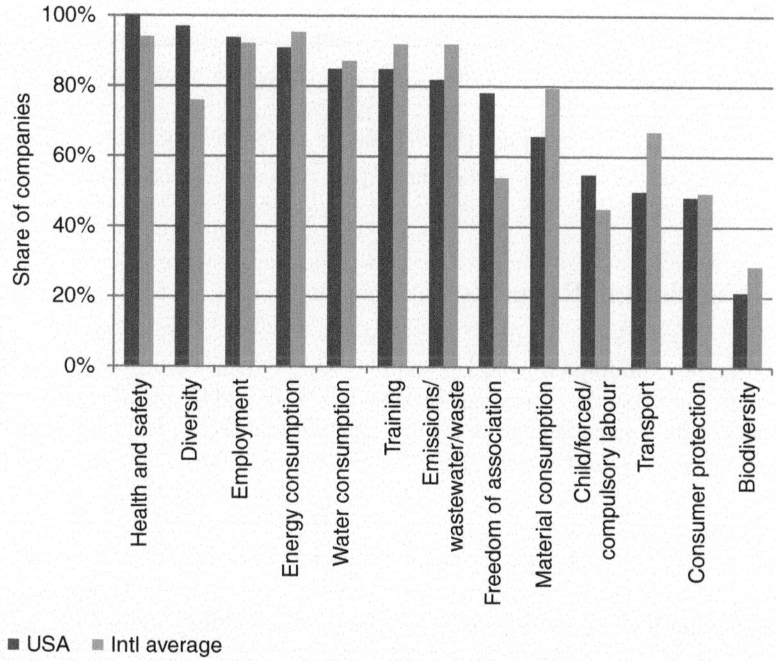

Fig. 14.8 Measured sustainability impacts (per cent; $n_{USA} = 31$, n_{INT} ranging from 425 to 454)

Based on the deliberation that companies are well advised to assess and measure the impacts and the success of their sustainability efforts to enable verifiable achievements, the implementation of corporate sustainability was also investigated in terms of what is measured by the companies. First, Fig. 14.8 points out that measuring sustainability impacts varies widely among the different sustainability issues. *Occupational health and safety* and *energy consumption*, for instance, are measured by a vast majority of companies whereas *biodiversity* is measured least. Second, the social issues *diversity/equal opportunity* and *freedom of association* are measured observably more often by the US companies than on international average; *material consumption* and *transport*, however, are measured less often on the national level. These results provide another indication – similar to the findings on managing sustainability issues (see Fig. 14.4) – that *material consumption* is perceived as not being of high relevance for the US companies while *energy consumption* currently is of high interest. Furthermore, these results underline the argument that the US companies surveyed pay considerable attention to social issues.

The companies were also asked if they measure the impact of their sustainability management on the company success or competitive advantage with regard to the drivers of business cases for sustainability: reputation, risks, revenue, efficiency, costs, innovation (products/processes) and business model innovation (see also

Fig. 14.5). Strikingly, the percentages of US companies that state they measure the impact of these subjects are sometimes slightly but consistently higher than the international average. The impact of reputation, for instance, is measured by 69 % of the US companies surveyed while 52 % of the companies on international average measure this impact. 28 % of the US companies and 27 % if the companies on the international level, however, measure the impact of business model innovation.

14.3 Concluding Remarks

In summary, the findings for the US sample, which mainly consists of very large companies, show that the companies surveyed know of and frequently apply numerous sustainability management tools such as *flexible working time*, *environmental management systems* and *sustainability reporting*. In addition, *corporate giving* and *corporate citizenship* are ranked high compared to the international average, which supports the argument that philanthropy is high on the agenda of US companies (Crane et al. 2014). The US companies, moreover, manage their stakeholder relationships in a more progressive and intensive way than do companies from the other countries surveyed for the Corporate Sustainability Barometer. Another result is that of all stakeholders and all countries, NGOs are assessed as most promoting by the US companies, whereas top management does not belong to the three most-promoting organisational units. This is an indication that sustainability is not necessarily incorporated on a strategic level and that corporate sustainability management may not be viewed strategically as core goal of the organisation but rather as an obligation. Innovation and expectations regarding revenue are considered less as drivers of a business case for sustainability meaning that in the US corporate sustainability is considered less as boosting the company's profit.

References

Blanchflower D, Freeman RB (1990) Going different ways: unionism in the U.S. and other advanced OECD countries. National Bureau of Economic Research working paper, no 3342. http://www.nber.org/papers/w3342. Accessed 15 Mar 2013

Bowen HR (1953) Social responsibilities of the businessman. Harper, New York

CNN Money (2012) FORTUNE 500 – our annual ranking of America's largest corporations. http://money.cnn.com/magazines/fortune/fortune500/index.html. Accessed 7 Jul 2012

Crane A, Matten D, Spence L (2014) Corporate social responsibility: in global context. In: Crane A, Matten D, Spence L (eds) Corporate social responsibility: readings and cases in global context, 2nd edn. Routledge, London, pp 3–20

Creyts J, Granade HC, Ostrowski KJ (2010) U.S. energy savings: opportunities and challenges. McKinsey Quart. http://www.mckinseyquarterly.com/US_energy_savings_Opportunities_and_challenges_2511. Accessed 5 Jun 2013

Epstein EM (1989) Business ethics, corporate good citizenship and the corporate social policy process: a view from the United States. J Bus Ethics 8(8):583–595

Goetsch DL (2005) Occupational safety and health for technologists, engineers, and managers, 5th edn. Pearson/Prentice Hall, London/Englewood Cliffs, pp 1–18
Hahn T, Scheermesser M (2006) Approaches to corporate sustainability among German companies. Corp Soc Responsib Environ Manag 13(3):150–165
Hansen EG (2010) Responsible leadership systems: an empirical analysis of integrating corporate responsibility into leadership systems. Gabler, Wiesbaden
Hess D, Rogovsky N, Dunfee TW (2002) The next wave of corporate community involvement. Corporate social initiatives. Calif Manag Rev 44(2):110–125
Kendall K (2006) Occupational safety and health in five countries. Technical report 8. National Occupational Health and Safety Advisory Committee (NOHSAC), Wellington
Kiron D, Kruschwitz N, Haanaes K, Reeves M, Goh E (2013) The innovation bottom line. How companies that see sustainability as both a necessity and an opportunity, and change their business models in response, are finding success. MIT Sloan Management Review/Boston Consulting Group, Boston
Kruger J (2005) From SO2 to greenhouse gases: trends and events shaping future emissions trading programs in the U.S. Resources for the future. Discussion paper, no 05–20, Washington, DC
LaDou J (ed) (1986) Introduction to occupational health and safety. National Safety Council, Chicago
Loew T, Ankele K, Braun S, Clausen J (2004) Significance of the CSR debate for sustainability and the requirements for companies, English summary. Berlin/Münster. http://www.ioew.de/uploads/tx_ukioewdb/future-IOEW_CSR-Study_Summary.pdf. Accessed 15 Mar 2013
Myers D, Pitkin J (2010) Assimilation today: new evidence shows the latest immigrants to America are following in our history's footsteps. Center for American Progress, Washington, DC
Schaltegger S (2011) Sustainability as a driver for corporate economic success. Consequences for the development of sustainability management control. Soc Econ 33(1):15–28
Schaltegger S, Burritt R (2005) Corporate sustainability. In: Folmer H, Tietenberg T (eds) The international yearbook of environmental and resource economics 2005/2006. Edward Elgar, Cheltenham, pp 185–222
Schaltegger S, Herzig C, Kleiber O, Müller J (2002) Sustainability management in business enterprises: concepts and instruments for sustainable organisation development. Bundesministerium für Umwelt, Naturschutz und Reaktorsicherheit (BMU)/Centre for Sustainability Management (CSM), Berlin/Lüneburg
Schaltegger S, Lüdeke-Freund F, Hansen EG (2012) Business cases for sustainability. The role of business model innovation for corporate sustainability. Int J Innov Sustain Dev 6(2):95–119
Schaltegger S, Harms D, Hörisch J, Windolph SE, Burritt R, Carter A, Truran S, Crutzen N, Ben Rhouma A, Csutora M, Tabi A, Kokubu K, Kitada H, Haider MB, Kim J-D, Lee K-H, Moneva JM, Ortas E, Álvarez-Etxeberria I, Daub C-H, Schmidt J, Herzig C, Morelli J (2013) International corporate sustainability barometer: a comparative study of 11 countries. Centre for Sustainability Management, Lüneburg
Tencati A, Perrini F, Pogutz S (2004) New tools to foster corporate socially responsible behavior. J Bus Ethics 53(1–2):173–190
U.S. Census Bureau (2013) International data base. http://www.census.gov/idb/ranks.html. Accessed 15 Mar 2013
Vigdor J (2008) Measuring immigrant assimilation in the United States – introduction. Civic report, no 53. Manhattan Institute for Policy Research. http://www.manhattan-institute.org/pdf/cr_53.pdf. Accessed 15 Mar 2013
World Bank (2013) GDP ranking, world development indicators. http://data.worldbank.org/data-catalog/GDP-ranking-table. Accessed 15 Mar 2013
World Commission on Environment and Development (WCED) (1987) Report of the World Commission on Environment and Development. Our common future. http://www.un-documents.net/wced-ocf.htm. Accessed 15 Mar 2013
Wright EO, Rogers J (2010) Chapter 2: What kind of a society is "American Society"? In: American society: how it really works. WW Norton, New York. http://www.ssc.wisc.edu/~wright/ContemporaryAmericanSociety/Chapter%202%20--%20what%20kind%20of%20society%20--%20Norton%20August.pdf. Accessed 15 Mar 2013

Part III
Patterns and Conclusion

Chapter 15
General Patterns and Conclusions

Stefan Schaltegger and Dorli Harms

Abstract Do global patterns of corporate sustainability management exist or are national differences reflected in substantially different corporate practices? This chapter reviews the literature supporting these two lines of argumentation. By taking a 'bird's view' of the overall patterns of sustainability management, we examine whether similarities or differences are predominant on the basis of the International Corporate Sustainability Barometer data. The key findings are that similarities predominate for the overall pattern but some national differences are found when examining the results in detail.

15.1 Introduction

Are corporate sustainability management practices expressed in globally similar patterns or are national differences reflected in substantially different corporate practices? This chapter takes a 'bird's view' of overall patterns of sustainability management and examines whether similarities or differences are predominant.

Two lines of argumentation can be distinguished. Firstly, large international companies worldwide could show many parallels in their sustainability management as many sustainability challenges, such as climate change, loss of biodiversity, poverty or population growth, have a global character. Large companies furthermore are publically exposed in international markets and media.

S. Schaltegger (✉) • D. Harms
Centre for Sustainability Management (CSM), Leuphana University, Lüneburg, Germany
e-mail: schaltegger@uni.leuphana.de; dharms@uni.leuphana.de

S. Schaltegger et al. (eds.), *Corporate Sustainability in International Comparison*, Eco-Efficiency in Industry and Science 31, DOI 10.1007/978-3-319-06227-3_15,
© Springer International Publishing Switzerland 2014

Secondly, different national regulations, cultures, markets and natural conditions could be expected to substantially influence companies, thus resulting in large national differences in corporate sustainability management practices.

This investigation is based on a comparison of the average national results for each surveyed item, whether the dispersion of the average scores are ‘close together’ or ‘far apart’, whether priorities or rankings are the same or largely different and whether the intensity levels of corporate sustainability management practices differ or not.

Overall, the analysis of the International Corporate Sustainability Barometer (ICSB) survey data – grouped according to the ‘Triple I’ approach: intention, integration and implementation – shows a broad global pattern for large companies in the surveyed countries. In all of the 11 economically developed countries, the management of corporate sustainability issues is strikingly similar. With few exceptions the average results for all countries reflect similar priorities, choices and activities. A closer examination of the data, however, reveals important differences on a more detailed level. The international similarities are largest for the intentions of large companies whereas a much wider spread of national differences can be identified for the implementation of corporate sustainability. The chapter concludes with recommendations.

15.2 International Similarities or National Differences in Corporate Sustainability Management?

Why do companies manage sustainability? To what extent is corporate sustainability embedded in their core business and organisation? How is corporate sustainability operationalized? These three questions, which embrace the intention, integration and implementation of corporate sustainability, also referred to here as the ‘Triple I’ approach, guided the analysis of sustainability management practices of large companies. In 2012 a total of 468 companies from 11 economically developed countries around the world took part in the ICSB study, which was initiated and coordinated by the Centre for Sustainability Management in Lüneburg (Schaltegger et al. 2013).

The analysis of the state of the art and progress of sustainability management and corporate social responsibility (CSR) practice on the international level to identify overall patterns reveals two complementary lines of argument.

On the one hand, large companies could show close similarities in their sustainability management resulting from global challenges like climate change, biodiversity loss, poverty or population growth. On the other, different countries could have distinct regulatory frameworks, cultures and markets, which might create disparities in national corporate practices dealing with sustainability. These two lines of argument are explained in more detail in the following.

15.2.1 International Similarities in Corporate Sustainability Management

New institutional theory may help to explain international convergence among large companies as the result of normative pressures, mimetic processes and coercive isomorphism (DiMaggio and Powell 1983). Institutional pressures may help to explain or predict similarities of CSR and corporate sustainability practices (Matten and Moon 2008).

As large companies do business around the world, they may face similar sustainability challenges, international stakeholders and market conditions, potentially resulting in similar patterns of their sustainability management practices. In a globalised world it would not be surprising if current and future issues such as emissions and waste reduction, resource scarcity, and occupational health and safety were central to all large companies no matter in which country their headquarters are located. Sustainability topics, media broadcasts, international standards and stakeholders cross national borders. The effect of greenhouse emissions on climate change, for instance, is widely discussed in many countries with regard to how to best achieve effective reduction. Not just businesses but also international organisations, media, policy makers and societal groups articulate their concerns and stimulate the discussion how to counteract increasing greenhouse gas emissions worldwide (Wright et al. 2013).

However, managing corporate sustainability is not just focussed on addressing environmental, social or economic risks. Sustainability management can also provide opportunities for large international companies if they can create competitive advantages by offering sustainability-oriented products and services. If a company can effectively promote sustainability-oriented innovations (Paech 2007; Hansen et al. 2009; Hansen and Klewitz 2012), this may help to attract new customer groups and to enlarge market share (Schaltegger and Burritt 2005). Car sharing exemplifies such an innovation that contributes to sustainable development and provides new business opportunities for progressive companies (Tietze and Hansen 2013).

Similar practices by large companies may also derive from the expectations of international or internationally networked stakeholders. International authorities and organisations like the United Nations and nongovernmental organisations (NGOs) act globally. Consumer protection serves as another example of stakeholder demands with a global reach (Crane et al. 2014). In addition, standards and norms such as the ISO family of standards, the OECD Guidelines for Multinational Enterprises or the Sustainability Reporting Guidelines by the Global Reporting Initiative (GRI) are implemented worldwide.

15.2.2 *National Differences in Corporate Sustainability Management*

New institutional theory could also be used to explain national differences in reference to national characteristics such as the political system or the cultural dimension (DiMaggio and Powell 1983; Whitley 1997; Matten and Moon 2008). If national institutional pressures were stronger than international influences, then national historical, political, cultural or regulatory differences would be a source of identifiable national differences in corporate sustainability management.

Matten and Moon (2008) discuss differences in corporate sustainability between the United States of America (US) and Europe by distinguishing between explicit and implicit CSR. While explicit means that a company's commitment and endeavours in sustainability are considered to be voluntary and often strategic and CSR is undertaken at the company's discretion, implicit corporate practices are less deliberate and can be understood as a reaction to external requirements and obligations (Porter and Kramer 2006; Matten and Moon 2008). Matten and Moon (2008) classify countries based on an assessment of corporate practices and conclude that companies in the US tend to make their CSR engagement more explicit than companies in European countries.

Various researchers have examined the relationship between national culture and corporate sustainability practices or ethical decision-making in business (e.g. Vitell et al. 1993; Husted 2005; Vachon 2010). Vachon (2010) investigates Hofstede's (1994) four cultural dimensions (individualism, uncertainty avoidance, power distance and masculinity) with regard to corporate sustainability. His analysis reveals that individualism and uncertainty avoidance have an influence on corporate sustainable development. Companies from a country with a high degree of individualism and a low degree of uncertainty avoidance, such as the US, show a more pronounced willingness to engage in voluntary sustainability practice but they are less likely to engage formally in corporate sustainability than countries like Japan that are rather characterised as being less individualistic and characterised by a higher degree of uncertainty avoidance (Vitell et al. 1993; Vachon 2010).

15.3 Analytical Approach

Our analysis of the similarities and differences among corporate sustainability practices in different countries focuses on the spread, patterns and the intensity level of sustainability management.

The *spread* among the average national results focuses on a single aspect or possible response and looks at the extent to which the national averages differ. For example, for the question how a company would assess the impact of different stakeholders on the implementation of corporate sustainability, do the companies in all 11 surveyed countries provide the same response on average (e.g. that NGOs are very supportive) or is there a wide spread of responses (e.g. some countries consider NGOs to be exceedingly supportive whereas others consider NGOs to be rather hindering)?

The *pattern* of responses refers to all possible responses or aspects of a question and compares the national averages. For example, do all national average responses indicate NGOs to be most supportive, customers to be of medium support and banks to be least supportive or are there significant differences?

The third indicator used in this cross-national analysis is *intensity level*. The intensity of corporate sustainability management practices is reflected by the respective levels of patterns among countries. For example, for the question to which extent companies expect positive impacts from the implementation of corporate sustainability, large companies in most of the countries surveyed score corporate reputation highest and revenue increase lowest. However, the intensity level is different as more Japanese companies than Belgian companies responded in the affirmative.

The 'Triple I' approach provides a framework for analysing these findings on a broader level. Intention, integration and implementation can be understood as three successive clusters, with each succeeding cluster involving stronger forms of engagement than the preceding one: Managing sustainability begins with a company's motivation for sustainability, its commitment, visions, missions and strategies for dealing with different environmental, social and economic aspects of sustainability. This is referred to as *intention* in the ICSB study (e.g. Bansal and Roth 2000; Epstein 2008; Babiak and Trendafilova 2011).

The second cluster, *integration*, goes one step further and examines the extent to which corporate sustainability is embedded in core business in a way that increases competitiveness as well as organisational involvement to ensure full organisational commitment (e.g. Porter and Kramer 2006; Schaltegger et al. 2012, 2014).

The third cluster, *implementation,* looks at the operationalization of corporate sustainability, how stakeholder relationships are managed, what sustainability management tools are known and applied and what issues and impacts are measured (e.g. European Commission 2004; Krick et al. 2005, Schaltegger et al. 2002).

Against this background of the two lines of argumentation, proposing either large similarities or large national differences, the following sections discuss the key findings with regard to spread, patterns and intensity level. The aim is to draw an overall comparative picture rather than to go into the details of specific questions, responses or national particularities as the latter have been dealt with in the previous country-specific chapters.

15.4 Key Findings

The key findings of the analysis of whether global similarities or national differences are prevailing are discussed with regard to spread (15.4.1), patterns (15.4.2) and intensity level (15.4.3).

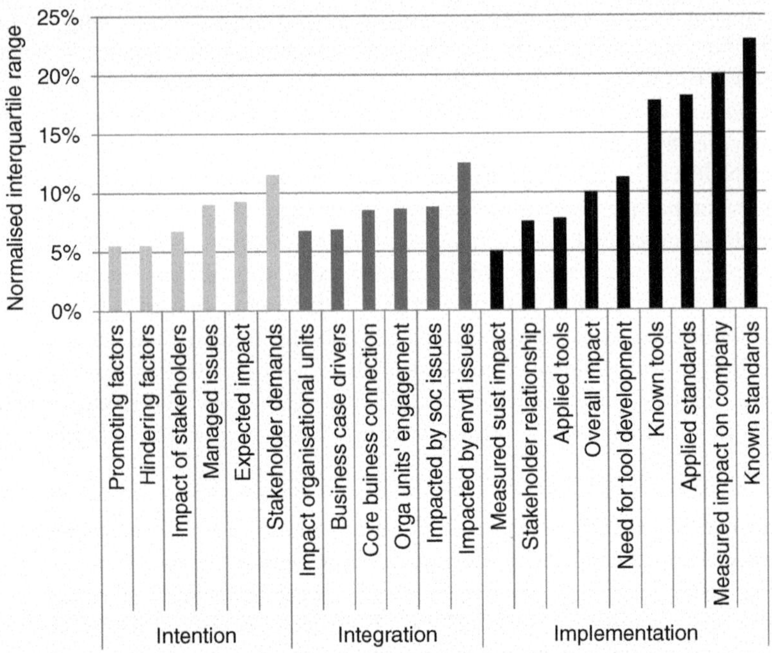

Fig. 15.1 Spread among national average results measured with normalised interquartile range (per cent; n_{INTL} ranging from 393 to 468)

15.4.1 The Spread Among National Averages

Analysing the *spread* of the national average results shows, with only few exceptions, large similarities for most investigated questions. The normalised interquartile range (see Fig. 15.1), which measures how far apart the most extreme national average results are from each other, shows that priorities, assessments and approaches to managing corporate sustainability are fairly similar among the countries. Nearly all of the investigated 21 aspects show a spread of less than 15 %. The four exceptions are known tools and applied standards (each about 18 %), the measured impact on the company (20 %) and known standards (23 %).

Grouping the interquartile range according to the 'Triple I' approach of intention, integration and implementation provides the basis for a second observation (Fig. 15.1). The results very clearly show smaller spreads for intention than for integration and for the implementation of corporate sustainability.

The growing spread of national average results from intention to integration and implementation (i.e. from left to right in the three groups in Fig. 15.1) corresponds firstly to the increasing resources required and secondly to the typical process of introducing and establishing sustainability management in a company.

Intentions are expressed in, for example, visions, policies or goals. These can be produced and publicised at relatively low cost and with little time or effort required. However, to score high in the category integration requires actually making sustainability part of core business activities and involving all departments in this endeavour. Activities in this group cause higher costs as they require more time and a broader involvement of staff. Implementation poses even more challenges and requires more resources such as acquiring knowledge and applying specific sustainability management tools, managing stakeholder relationships by intensive exchange and empowerment activities or measuring the effects and the efficiency of implementation activities of sustainability management.

The fact that the spread among national differences grows from intention to integration and then to implementation not only reflects the higher costs of the latter stages but also reflects maturity of how well sustainability management is implemented in companies. The basic stages intention, organisational integration and concrete implementation are also found in regard to standards on sustainability, environmental and social management systems such as ISO 26000, ISO 14001, EMAS or SA 8000.

Overall, the analysis of the spreads among national averages as well as the spreads at the intention, integration and implementation stages shows large similarities. The priorities, assessments and activities among large companies are thus more a global phenomenon than one characterised by national differences. However, the spreads in a stage increase with more advanced implementation activities and higher resource requirements. This means that large companies have on average not proceeded in every country to the same extent in establishing sustainability management practices and in allocating resources to corporate sustainability implementation. This shows that in spite of overall similarities, national differences in the progress and maturity of sustainability management practices do exist. The next section therefore investigates national patterns of corporate sustainability management.

15.4.2 National Patterns in Comparison

Patterns of responses address the national ranking of responses to all possible aspects of a single question. Are they similar or are there large differences among the countries surveyed? The three lines in Fig. 15.2 illustrate examples of patterns for the question which stakeholders are considered to be the five most and the five least supportive. The upper line links the responses for Japan and the lower line the responses for Belgium. Both patterns are similar, even for the assessment of the role of communities, which is slightly different from the overall international average pattern for all countries (the middle line).

Comparable to the findings on the spread of responses for each individual item (Sect. 15.1), the findings for the patterns show that the similarities among the countries are predominant rather than national differences.

For most questions the pattern lines for countries run in parallel rather than cross each other. For instance, the intention for corporate sustainability is mainly characterised by a societal orientation associated with safeguarding reputation

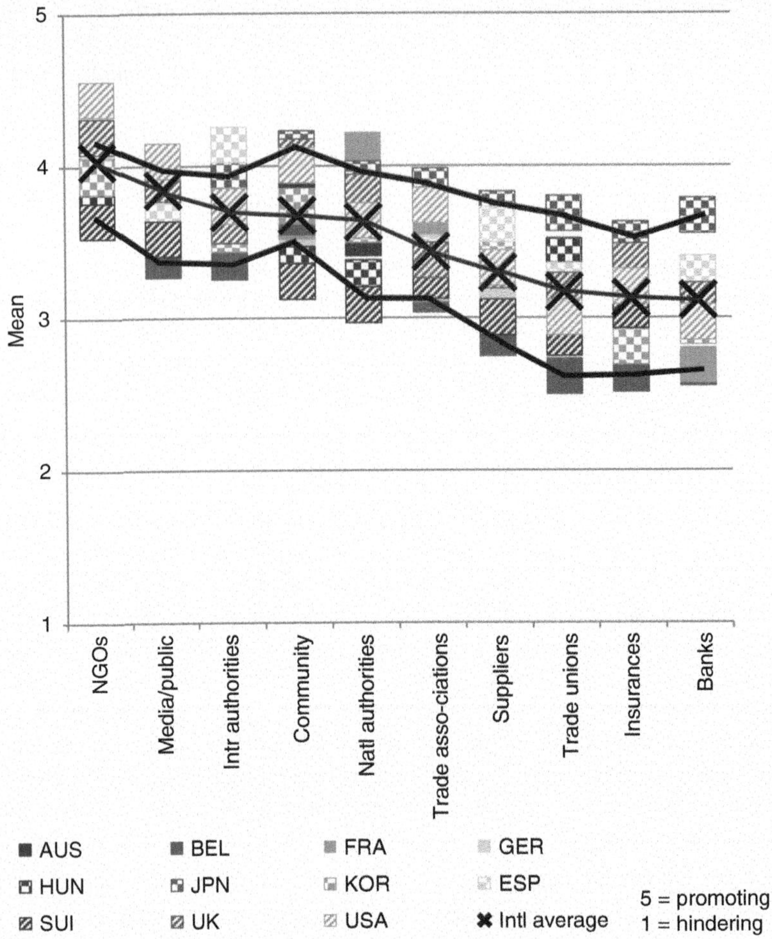

Fig. 15.2 Impact of external stakeholders on corporate sustainability, single countries may not be visible due to overlaps (n_{INTL} ranging from 393 to 450)

and securing legitimacy whilst market orientation in corporate sustainability management ranks behind. The pattern that societal stakeholders are perceived as the strongest supporters of corporate sustainability management whereas market-oriented stakeholders are perceived as least promoting characterises all countries (see Fig. 15.2; also Schaltegger et al. 2013, p. 19).

Another example showing a similar pattern is for the question for which environmental and social issues stakeholders expect a company to engage. Health and safety as well as employment are predominant issues in all surveyed countries whereas biodiversity and transport are less strongly addressed. This matches the pattern of sustainability issues which companies manage and where only few differences in national patterns exist (e.g. water is assessed much higher in Australia than in other countries).

Overall, the patterns of responses are characterised by substantially more similarities than national differences. This, however, does not reveal the weight generally given to sustainability as a topic and the intensity with which sustainability management is implemented.

15.4.3 Intensity Levels of Sustainability Management

The third indicator, the *intensity level* of managing corporate sustainability, reflects national differences with regard to the extent a particular aspect of sustainability management is practised. For example, Fig. 15.2 shows that Japanese companies (top line) consistently rank the impact of external stakeholders on corporate sustainability higher than the international average (middle line) and companies in Belgium (bottom line): the intensity level of this aspect of sustainability management is thus higher in Japan than in the international average or in Belgium.

Figure 15.3 illustrates another finding relating to the intensity of managing stakeholder relationships. Ranging from more passive forms such observing and informing stakeholders to more proactive ones such as empowerment or delegating decision making (modified from Krick et al. 2005), differences in the intensity level can be identified, for example, between the United States (upper line) and Switzerland (lower line). Large companies in the United States (US) use on average all forms of managing stakeholder relationships more intensively than large companies in Switzerland. Figure 15.3 (see also Schaltegger et al. 2013, p. 32) moreover shows that Korean companies, as in the US, have on average a high intensity pattern in managing their stakeholder relationships more closely than their international peers. High intensity levels in sustainability management support the argument of more explicit corporate practice.

These results support the analysis of Matten and Moon (2008) with regard to the US. However, the high intensity levels for Japan and Korea indicate that these countries use a more explicit sustainability management and thus do not support their analysis in this regard. These differing results may reflect recent developments. Matten and Moon (2008) themselves mention in their analysis of implicit and explicit CSR that a tendency exists also in countries with an implicit CSR to more often signal explicit engagement.

The theoretical literature on similarities or national differences in sustainability management shows national differences exist with regard to intensity levels.

Analysing the overall intensity levels reveals that intensity levels do indeed vary substantially, i.e. some countries are for all possible responses on one question in the top group whereas other countries are placed at the lower intensity end for all or nearly all responses. These differences are, however, only partially consistent with earlier research differentiating between implicit and explicit CSR.

Overall the international comparison based on the ICSB survey data reveals that Japanese companies often show intensity levels above the international average. On

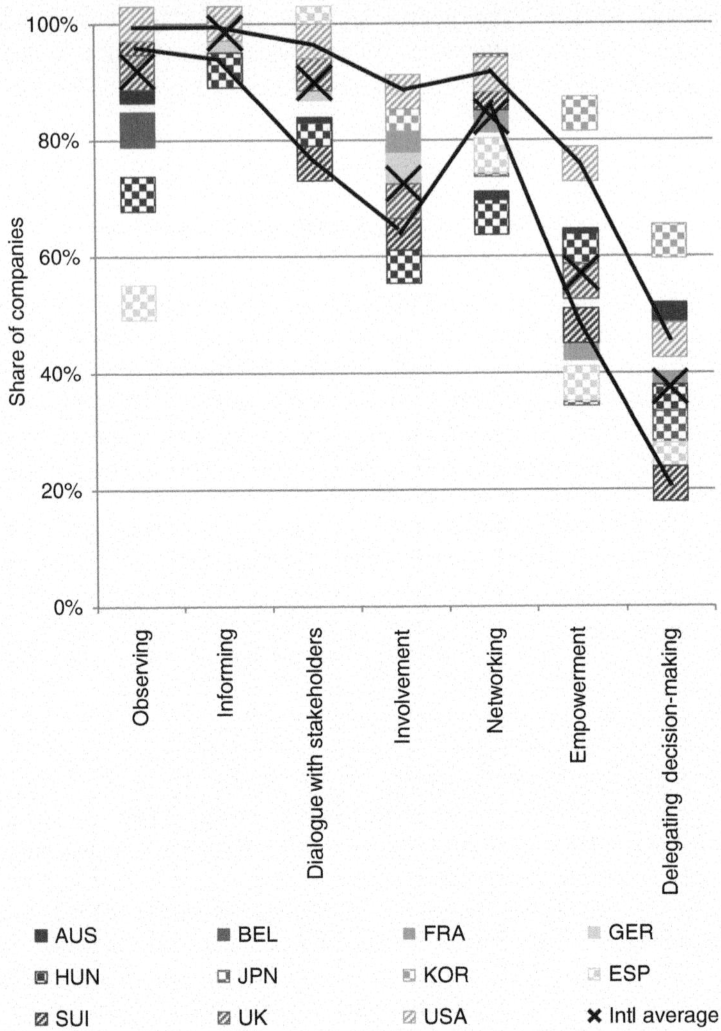

Fig. 15.3 Management of stakeholder relationships 438 to 458 (n_{INTL} ranging from 438 to 458)

the contrary, Switzerland and Belgium often show low intensity values. In the case of Switzerland (see Chap. 12) this may be a result of the large share of financial organisations in the sample, which on average emphasise sustainability management less than manufacturing industries. In the case of Belgium (see Chap. 5) the smaller company size on average may explain part of the deviation from the international intensity level. Both possible explanations, the influence of industry and company size, could not be tested here and may encourage further research.

15.5 Concluding Remarks

With the analysis of similarities and differences of corporate sustainability at an international level, this chapter extends the country-specific analyses presented earlier in this edited volume. Three measures were applied to analyse global similarities and national differences in corporate sustainability management practices: the spread of national results for specific responses, the patterns of responses to one question and the intensity levels.

The comparative analysis based on the first two measures, the spread and the pattern of national average results, reveals overall that the sustainability management of large companies has strikingly much in common in all surveyed countries. This justifies the interpretation that sustainability among large companies in economically developed countries is to a large extent a global phenomenon. This result is particularly pronounced for the supportive role of societal stakeholders, the orientation towards securing legitimacy, the relevance allocated to different sustainability issues and the involvement of different organisational units.

However, evidence was also identified for national differences in sustainability management practices. The increasing spread across intention, integration and implementation and the differences among the national intensity levels for many sustainability issues indicate that the progress achieved in implementing sustainability management is not the same in every country and that the large companies in the surveyed countries do not invest the same amount of resources for the implementation of corporate sustainability.

In light of the results of this ICSB study, further research is needed to investigate how the identified similarities and differences for the spread, patterns and intensity levels can be explained, why in some cases national differences exist whereas in others global patterns prevail and how practices change over time. Further investigations are also needed to examine the influence of differences among industries as well as between large and small companies.

For practitioners the comparative findings of the ICSB study presented in this edited volume can provide a valuable basis for designing and further developing their own sustainability management practices to better deal with global and national challenges and opportunities. The examination of the intention, integration and implementation of corporate sustainability practices reveals the potential for improvements to enhance corporate contributions for the sustainable development of the company as well as of the economy, environment and society.

References

Babiak K, Trendafilova S (2011) CSR and environmental responsibility: motives and pressures to adopt green management practices. Corp Soc Responsib Environ Manag 18(1):11–24

Bansal P, Roth K (2000) Why companies go green. A model of ecological responsiveness. Acad Manag J 43(4):717–736

Crane A, Matten D, Spence L (2014) Corporate social responsibility: in a global context. In: Crane A, Matten D, Spence L (eds) Corporate social responsibility: readings and cases in global context, 2nd edn. Routledge, London, pp 3–20

DiMaggio PJ, Powell WW (1983) The iron cage revisited: institutional isomorphism and collective rationality in organizational fields. Am Sociol Rev 48(2):147–160

Epstein MJ (2008) Making sustainability work. Best practices in managing and measuring corporate social, environmental, and economic impacts. Greenleaf, Sheffield

European Commission (2004) ABC of the main instruments of corporate social responsibility. http://www.coc-runder-tisch.de/news/news_juni_2005/csr_abc.pdf. Accessed 31 Oct 2013

Hansen EG, Klewitz J (2012) Publicly mediated inter-organizational networks: a solution for sustainability-oriented innovation in SMEs? In: Wagner M (ed) Entrepreneurship, innovation and sustainability. Greenleaf, Sheffield, pp 254–278

Hansen EG, Grosse-Dunker F, Reichwald R (2009) Sustainability innovation cube – a framework to evaluate sustainability-oriented innovations. Int J Innov Manag 13(4):683–713

Hofstede G (1994) Management scientists are human. Manag Sci 40(1):4–13

Husted BW (2005) Culture and ecology: a cross-national study of the determinants of environmental sustainability. Manag Int Rev 45(3):349–371

Krick T, Forstater M, Monaghan P, Sillanpää M (2005) Stakeholder engagement manual. In: The practitioner's handbook on stakeholder engagement, vol 2. AccountAbility, London

Matten D, Moon J (2008) 'Implicit' and 'explicit' CSR: a conceptual framework for comparative understanding of corporate social responsibility. Acad Manag Rev 33(2):404–424

Paech N (2007) Directional certainty in sustainability-oriented innovation management. In: Lehmann-Waffenschmidt M (ed) Innovations towards sustainability. Conditions and consequences. Physica Verlag, Heidelberg/New York, pp 121–140

Porter ME, Kramer MR (2006) Strategy and society. The link between competitive advantage and corporate social responsibility. Harv Bus Rev 84(12):78–92

Schaltegger S, Burritt R (2005) Corporate sustainability. In: Folmer H, Tietenberg T (eds) International yearbook of environmental and resource economics. Edward Elgar, Cheltenham, pp 245–289

Schaltegger S, Herzig C, Kleiber C, Müller J (2002) Sustainability management in business enterprises, concepts and instruments for sustainable organisation development. The Federal Ministry for the Environment, Nature conservation and Nuclear Safety (BMU)/Federation of German Industries (BDI), Bonn/Berlin

Schaltegger S, Lüdeke-Freund F, Hansen E (2012) Business cases for sustainability. The role of business model innovation for corporate sustainability. Int J Innov Sustain Dev 6(2):95–119

Schaltegger S, Harms D, Hörisch J, Windolph SE, Burritt R, Carter A, Truran S, Crutzen N, Ben Rhouma A, Csutora M, Tabi A, Kokubu K, Kitada H, Haider MB, Kim J-D, Lee K-H, Moneva JM, Ortas E, Álvarez-Etxeberria I, Daub C-H, Schmidt J, Herzig C, Morelli J (2013) International corporate sustainability barometer: a comparative study of 11 countries. Centre for Sustainability Management, Lüneburg

Schaltegger S, Harms D, Windolph SE, Hörisch J (2014) Involving corporate functions: who contributes to sustainable development? Sustainability 6(5):3064–3085

Tietze F, Hansen EG (2013) To own or to use: how product service systems impact firms' innovation behaviour. Eur Financ Rev Aug–Sept:53–56

Vachon S (2010) International operations and sustainable development: should national culture matter? Sustain Dev 18(6):350–361

Vitell SJ, Nwachukwu SL, Barnes JH (1993) The effects of culture on ethical decision-making: an application of Hofstede's typology. J Bus Ethics 12(10):753–760

Whitley R (1997) Business systems. In: Sorge A, Warner M (eds) The IEBM handbook of organizational behavior. International Thomson Business Press, London, pp 173–186

Wright C, Nyberg D, de Cock C, Whiteman G (2013) Future imaginings: organizing in response to climate change. Organization 20(5):647–658

Index

S. Schaltegger et al. (eds.), *Corporate Sustainability in International Comparison*, Eco-Efficiency in Industry and Science 31, DOI 10.1007/978-3-319-06227-3,
© Springer International Publishing Switzerland 2014

The manufacturer's authorised representative in the EU is Springer Nature Customer Service Centre GmbH, Europaplatz 3, 69115 Heidelberg, Germany. If you have any concerns regarding our products, please contact ProductSafety@springernature.com

Printed and bound by CPI Group (UK) Ltd, Croydon, CR0 4YY

25/09/2026

02222841-0006